APPLICATIONS MANUAL

ENERGY AUDITS AND SURVEYS

AM5 : 1991

The Chartered Institution of Building Services Engineers
Delta House, 222 Balham High Road, London SW12 9BS

697
APP

ISBN 0 900953 48 9

Typeset and printed in Great Britain by Unwin Brothers Ltd, Old Woking, Surrey

Foreword

This *Applications Manual* has been produced through a collaboration between the Chartered Institution of Building Services Engineers and the Building Research Energy Conservation Unit (BRECSU), to encourage more organisations to undertake energy surveys, and, ultimately, to implement energy saving measures targetted at a total of £30M per annum.

The purpose of the *Manual* is to assist in the preparation of specifications, commissioning and undertaking surveys, and the potential audience includes energy managers, consultants and building users.

The *Manual* provides advice on how to identify and specify the requirements of an audit and site investigation; and how to carry out an appropriate level of study and implement recommendations. Each stage, from preliminary data analysis to final monitoring, is covered. The methods described are suitable not only for in-house studies but also form the basis of good practice for independent specialists. The *Manual* will have widespread application to buildings of all sizes and end uses throughout the public, commercial and industrial sectors.

The *Manual* was drafted by NIFES Consulting Group under contract to BRECSU, and the project was supervised by a small task group of volunteers nominated by CIBSE. This collaborative work is administered by a CIBSE/BRECSU Joint Steering Committee, under the overall supervision of the Institution's Technical Publications Committee. The aim of the collaboration is to accelerate the production of publications on energy-related topics in order to realise quickly the improvements in energy efficiency which should result from the application of the guidance presented. The collaboration is possible due to the generous support of the Department of Energy through its Energy Efficiency Office.

D J Stokoe
Chairman, Technical Publications Committee

Energy Audits and Surveys Task Group

N P Howard *Chairman*
P Jones
E M McKay
L McKay
P Mayo
B W Copping *Secretary*

Contract Author

G E F Read

Co-ordinating Editor

B W Copping

Energy Efficiency Office
DEPARTMENT OF ENERGY

Publications Secretary

K J Butcher

The voluntary input of personnel from the following organisations is gratefully acknowledged: Building Energy Solutions, John Lewis Partnership, Oscar Faber Consulting Engineers, Property Services Agency

Cover illustrations: *Lower right* Building Research Establishment (Crown copyright 1991); *Upper left* Shearson Lehman trading floor, Broadgate, London (courtesy Thorn Lighting Ltd); *Lower left* Energy survey in progress (courtesy NIFES Ltd)

Contents

List of tables

List of figures

Energy audits and surveys

1 Introduction

1.1 Background

Energy audits and surveys are investigations of site energy use aimed at identifying measures for cost and energy savings. They are an essential part of the effective control of energy costs and should be undertaken regularly, typically every three to five years.

All organisations, regardless of size, need to satisfy themselves that they are getting good value from their expenditure on energy. Audits and surveys, combined with routine monitoring, provide the vital information needed to ensure that energy is managed properly.

Most organisations already recognise that efficient energy use can not only reduce operating costs but also produce important environmental benefits. We all now have a responsibility to conserve natural resources and to reduce the harmful emissions that the combustion of fossil fuels can produce. Many organisations find that a responsible environmental attitude plays an important part in promoting their activities.

Nevertheless, the financial rewards are normally the underlying reason for implementing energy efficiency measures. Every pound saved on energy costs could increase profits directly by the same amount. Even organisations operating efficiently can expect a survey to identify potential energy cost savings of up to 10%. Cost savings of over 60% have been achieved at sites where the potential had not previously been recognised. Such savings are usually obtained with payback periods of less than three years. An Energy Efficiency Office study[1] of a sample of 4331 energy surveys showed that average savings of 21% of each site's energy bill were identified. The average payback period for recommendations was 1.5 years, providing an excellent opportunity for investment.

With proper guidance, organisations throughout all sectors can benefit, whether studies are carried out in-house or with specialist outside assistance. Accordingly, this *Manual* has been developed in the form of a guide to good practice on how to:

- adopt an appropriate strategy
- identify individual needs
- conduct an audit and survey
- assess the options
- proceed to implementation.

1.2 Aims of Manual

This document:

- assists in identifying the requirements for an audit or survey and in selecting the approach to be adopted
- encourages energy users to carry out their own energy audits, to the extent that they have in-house expertise, and to realise the value of what they can achieve themselves
- encourages the uptake of external energy audit services, builds confidence in the results of such services and encourages the implementation of recommendations
- assists users to assess the value of commissioning outside assistance, and to improve the briefing of consultants
- assists energy audit providers by establishing a framework of good practice
- assists all parties in maintaining effective communication throughout a project
- provides models for specifying the scope of comprehensive or concise energy audits and surveys.

1.3 Intended readership

The *Manual* is intended for anyone concerned with building energy use and, in particular, those directly responsible for the control of energy costs. It is recommended for the part-time energy manager and full-time specialist alike. This includes energy efficiency officers, engineering managers, energy consultants and, in organisations without an engineering capability, managers from other disciplines. Summaries at the end of each section provide an overview for quick reference by senior management.

1.4 Scope of Manual

This *Manual* has application throughout the public, commercial and industrial sectors, although there is no intention to cover industrial processes. It can be used for the full range of premises from extensive factory sites or office developments to single small buildings. Domestic properties are not specifically covered but some of the principles could still be applied to these with discretion.

The Manual recognises that water is often included in the scope of an energy audit or survey. Some outline guidance on surveying the purchase and use of water is included in Appendix A11.

1.5 Summary of section 1

* Site surveys should normally be carried out every three to five years.
* Energy efficiency provides important financial and environmental benefits.
* Potential energy cost savings of around 20% are typically identified. Savings as high as 60% have been found in some surveys and even efficient organisations could identify worthwhile savings of up to 10%.
* This Manual provides guidance on how to plan, specify and conduct energy audits and surveys and implement the recommended findings.

Reference for section 1

1 *Energy Efficiency Survey Scheme — An energy savings success* (London: Energy Efficiency Office) (November 1985)

2 Guidance for readers

2.1 Structure of Manual

The *Manual* is structured to lead readers from the initial planning stages through the audit and survey procedures to the final implementation of recommendations. Those intending to conduct a full in-house study themselves will benefit from reading all sections. Others, making more selective use of the *Manual*, are still advised to familiarise themselves with its full content.

The main purpose and target reader of each section are given in Table 2.1.

Table 2.1 Main purpose of each section of the *Manual*

Section		In-house auditor	Survey specialist	Purpose
1	Introduction	✓	✓	Background
2	Guidance	✓	✓	Definitions
3	Benefits	✓	—	Justification
4	Approach	✓	—	Methodology
5	Preliminary audit	✓	✓	Set priorities
6	Planning	✓	✓	Decide brief
7	Energy management	—	✓	Site survey
8	Energy supply	—	✓	Site survey
9	Energy conversion	—	✓	Site survey
10	Energy use	—	✓	Site survey
11	Findings	✓	✓	Report
12	Implementation	✓	—	Action
Appendices		✓	✓	Reference

2.2 Using the Manual

2.2.1 Energy users/in-house auditors

Most readers should be able to carry out preliminary audit procedures. Even those with limited knowledge of energy and building services are encouraged to interpret basic consumption data and establish the priorities for a site survey. By following the guidance given in this *Manual*, many will also be able to conduct an in-house survey which will identify at least some of the main opportunities for energy savings. They will certainly find it of help to take stock of the purchase and use of energy in their buildings. The potential value of such in-house studies should not be underestimated.

For those unable to allocate sufficient time or who do not have all the skills and resources for detailed survey work, the assistance of an outside specialist can be commissioned.

Prospective users of professional energy audit services should identify those tasks which are within the limit of their own capabilities and recognise where additional skills and resources are needed. The non-specialist auditor should then use the *Manual* either directly as a guide to good practice or indirectly as a checklist when commissioning outside consultants.

When an outside specialist is commissioned, a clear understanding must be established on both sides as to the objectives. With proper co-operation, the most effective use can be made of the combined skills and resources. A good brief is the key to the best use of consultants. Reference to the models for comprehensive and concise surveys included in the Appendices is strongly recommended.

2.2.2 Consultants/survey specialists

Providers of professional energy audit services should use the *Manual* as a framework for good practice. They should not confine their interest to the audit and survey procedures, where specialist technical knowledge is assumed. Initial agreement on the survey brief and, ultimately, implementation must also be given full attention.

Specialists will, of course, retain their own particular style of conducting an audit and survey, but the general principles set out in the *Manual* deal with the most common requirements and can be used as an *aide memoire.*

It must be stressed that the best results will come from a flexible and imaginative approach to the collection and interpretation of data. However, in departing from the procedures outlined, the reader should be satisfied that alternative methods are appropriate.

The *Manual* is not intended for use as a rigid specification in itself but again, the importance of establishing a good brief cannot be overstressed. This can ensure that the expectations from a project are understood and that the requirements can be fulfilled.

2.3 Definitions

Audits, surveys and monitoring are all key elements of a good energy management policy. This document is concerned primarily with audits and surveys. Figure 2.1 indicates their relative positions in a plan for reducing energy costs. Each of the principal stages is shown with a reference to the relevant section of the *Manual.*

2.3.1 Energy audit

An energy audit is a study to establish the quantity and cost of each form of energy input to a building, site or organisation over a given period. This could include evaluation of the end use and would include any on-site generation.

A simple preliminary audit can be performed with little specialist knowledge about energy. It will deal primarily with energy bills, and therefore does not involve a detailed site investigation.

An improved audit is often possible where site records of sub-meter readings are kept. Some engineering knowledge is useful for checking the reliability of site data.

A full audit requires expertise to break down energy use on a service-by-service basis. The process normally involves

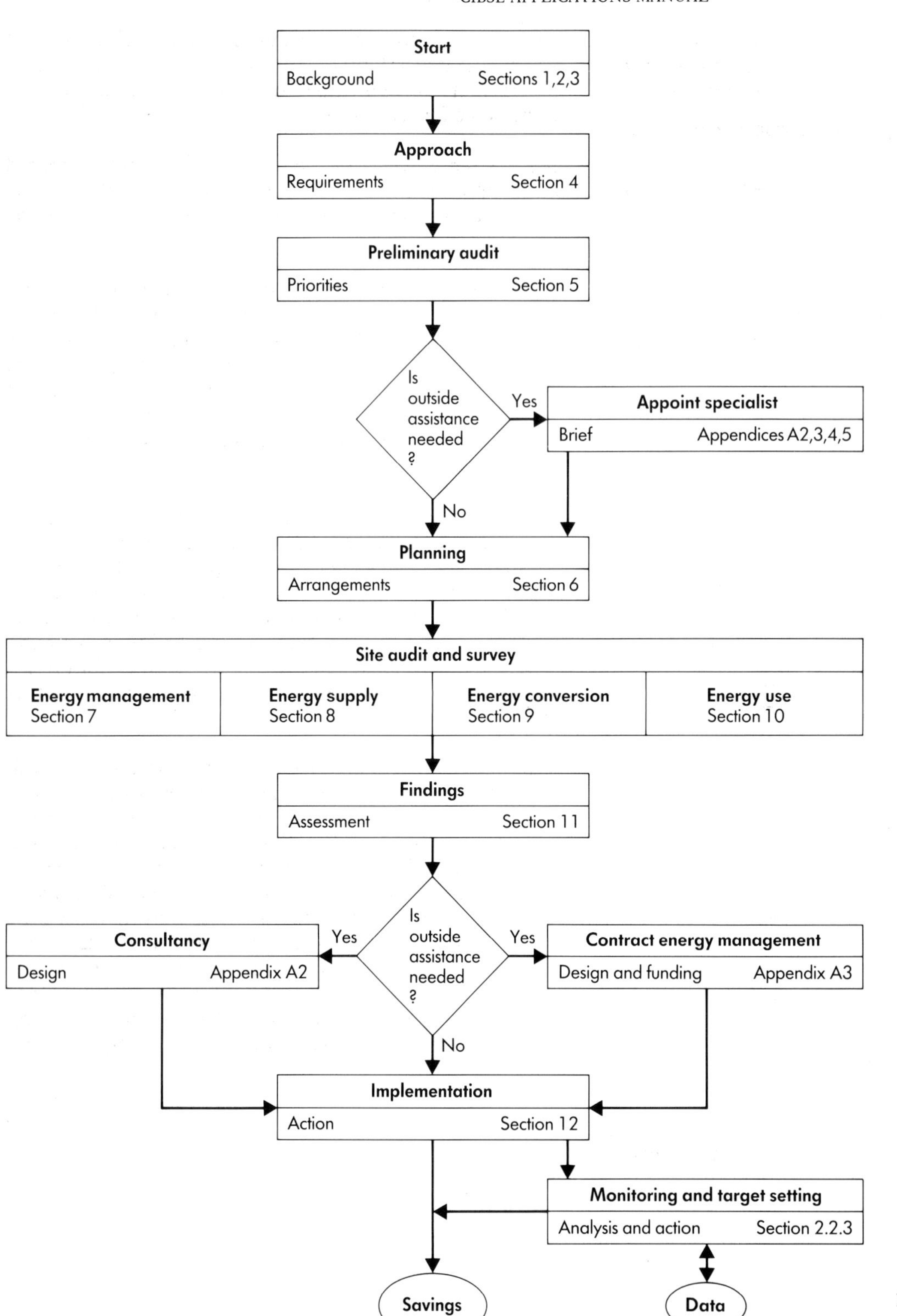

Figure 2.1 Flow chart for energy audit and survey

measurement, analysis or direct assessment of energy consumption to indicate the proportions attributable to heating, lighting, air conditioning or other major uses. Such information can only be obtained by performing a site survey.

Analogously to a financial audit, an energy audit provides management information to assist in decision making and will typically be based on the most recent financial year.

The results can be used to:

— determine the priorities for more detailed investigation

— justify investment in energy efficiency measures

— identify cases for direct action, e.g. choice of fuel type, change of fuel supply tariff

— raise the energy awareness of all staff by providing facts on energy use.

2.3.2 Energy survey

An energy survey is a technical investigation of the control and flow of energy in a site, building or process. It should aim to identify cost-effective energy saving measures. Surveys can vary in the range and depth of study but could include some or all of the following elements:

- development of an audit, supplemented in a comprehensive survey by measurement of principal energy flows and performance assessment of major plant
- examination of building fabric, services, controls and processes
- examination of energy supply and distribution arrangements
- consideration of occupancy, building use and environmental conditions and requirements
- analysis of energy performance in relation to building size, type, location, occupancy and climate with reference to performance indicators
- review of energy management policy and procedures including staff resources, monitoring and target setting, investment, planning and maintenance
- identification of opportunities for energy and cost savings with recommendations for action.

Firm recommendations can be produced following technical and financial appraisal of the available options.

A survey should be performed typically every 3–5 years, depending on the rate at which a site is changing, or following any major change in circumstances, including:

- change in building use or new tenants
- major refurbishment or development
- rationalisation leaving extensive unoccupied areas
- revised working practices or occupancy patterns
- substantial changes in fuel prices or availability
- significant upward movement in site energy consumption or costs.

It may sometimes be appropriate to perform a survey before major redevelopment to provide up-to-date information.

2.3.3 Energy monitoring

Energy monitoring involves the regular recording of energy consumption and cost and the principal variables, such as outside temperature and occupancy, which affect them. It allows essential information on energy performance to be provided at the right time and in a useful form to those responsible for its control.

Surveys can provide much of the basic information needed to set up a monitoring scheme. They are usually the best way to develop targets against which performance can be measured.

A site is normally best considered as a collection of distinct cost centres, which can be monitored both individually and collectively. These often equate with areas currently used for accounting purposes, but each cost centre must be capable of having its energy use metered. A study of energy use can allow targets to be set for improvements in energy performance in each area for which specific responsibility can be assigned.

Data analysis takes account of the effect which uncontrollable variables, such as outside temperature, may have on consumption. The effect of management action on the controllable variables, for example operating practice, can then be evaluated. A quantitative assessment can also be made of any savings achieved following the implementation of survey recommendations, and of the potential for further improvement. Longer-term trends might be revealed as more data are collected.

Typical monitoring periods are weeks or months, with quarterly and annual summaries. The periods chosen will reflect a balance between the value of energy being monitored at any point and the cost of collecting and analysing data. In most cases meter readings are collected by hand, along with information on the principal variables. The routine task of analysis can be greatly simplified by using microcomputer-based spreadsheet methods. Computer-based building energy management systems can extend this function to include automatic remote meter reading and data processing as part of an overall monitoring and control package. Results must still be interpreted and acted on.

2.4 Summary of section 2

- * An energy audit establishes the quantity and cost of each form of energy input and, as part of a site survey, the breakdown between end users.
- * An energy survey is a technical site investigation of selected aspects of the flow, control and management of energy to identify and evaluate opportunities for savings.
- * Surveys should be combined with regular monitoring of energy use.
- * This *Manual* provides guidance to energy users in all sectors, whether conducting their own audits and surveys or commissioning outside assistance.
- * The *Manual* should be used as the basis of good practice by those conducting energy audits and surveys.
- * A good brief is the key to successful use of outside consultants to carry out surveys. Model briefs for comprehensive and concise surveys are shown in Appendix A4 and Appendix A5 respectively.

3 Benefits of audits and surveys

The benefits that can result from an energy audit and survey fall into the following categories:

(*a*) Financial benefits which contribute to a reduction in operating costs or an increase in the profits of an organisation

(*b*) Operational benefits which assist the management of a building or site, improve the comfort, safety and productivity of its occupants or otherwise improve its general operation

(*c*) Environmental benefits; those which have an additional value outside the organisation by reducing environmental damage.

Each of the benefits is likely to be realised progressively and to have a cumulative effect. The principal benefits may become available immediately from no-cost measures, or could involve some period before a return on investment is achieved. Others might only be realised when longer-term plans are implemented.

Long-term measures are sometimes applied very cost effectively when phased in with other refurbishment, e.g. fabric insulation. Early savings can often provide the funding for investment in future projects as part of a phased programme of work. Implementation of part of a measure, e.g. selective fabric insulation or plant replacement, may help in this respect. With authority to reinvest savings, an energy manager is provided with a strong incentive to develop an energy saving programme.

3.1 Financial benefits

The financial benefits of the potential measures identified must be assessed against their cost of implementation. It is usual to group the available options according to cost, so that they may be considered at the appropriate level in an organisation.

Financial benefits can be realised in several ways, not necessarily dependent on the level of investment:

— reduced expenditure on energy, by reducing consumption or changing tariff or fuel type

— reduced maintenance costs following improved utilisation of plant and reduction in running hours

— savings in other costs, e.g. water charges, where demand is reduced

— reduced capital expenditure, e.g. where increased efficiency avoids the need for additional plant capacity

— more productive use of labour where measures release staff for other duties, e.g. boilerhouse automation

— increased productivity where working conditions are improved.

3.1.1 No-cost/low-cost measures

Measures can normally be identified which require minimal investment and can often be implemented without further study. These include:

— general good housekeeping; 'switch-off' campaigns; avoiding wasteful practices

— adjustment of existing controls to match actual requirements of occupancy

— change of fuel purchasing tariffs, or correct selection of fuel where there is a dual-fuel facility

— rescheduling of activities; planning to take advantage of tariff structures; changing the use of building space

— small capital items such as installation of thermostats, timeswitches and sections of pipework insulation or draughtproofing.

3.1.2 Medium-cost measures

These measures may still require little or no further design or study, but involve expenditure on works and will consequently take longer to implement. A convenient range of capital cost, e.g. £500 to £5000 per measure, can be defined to suit the requirements of any particular organisation.

Common investment measures include:

— installation of new or replacement controls for heating, cooling or lighting which not only control energy costs but could also improve comfort and increase productivity

— insulation or refurbishment of roofs, walls, windows and floors to reduce heat loss, prevent draughts or reduce solar gain as appropriate

— installation of additional metering and facilities to monitor energy use.

Depending on their precise value, such measures will sometimes fall into the low-cost or high-cost categories. Some measures can have a low marginal cost when incorporated into major refurbishment work, e.g. flat roof insulation.

3.1.3 High-cost measures

Measures in this category are expected to need further detailed study and design. Individual approval may have to be obtained at an executive level before implementation. Such measures could include:

— replacement or upgrading of plant and equipment, e.g. boilers, chillers, water heaters and luminaires, to improve efficiency and reduce energy, maintenance and other operating costs

— installation of a building energy management system to control and monitor building services and site energy flows

— major changes in the methods of generating heat or power, e.g. decentralisation of boiler plant or introduction of a combined heat and power scheme.

3.2 Operational benefits

In addition to direct cost benefits, further management benefits can arise in the operation of a building or site. Ultimately, these may well have financial implications. The information made available to management on energy costs and use could in itself be found invaluable in planning and decision making. Measures can also lead to improved working practices or conditions.

3.2.1 Management information

Information becomes available as an energy audit and survey progresses. It may be used to decide on immediate action or for longer-term planning. Information could take the form of:

— a statement of total expenditure on energy. Comparison with other operating costs will assist when assessing the case for investment in energy-saving measures.

— an assessment of energy performance, e.g. by relating consumption to the building area or volume. Comparison with performance indicators or yardsticks for similar types of building can be useful in establishing priorities for action.

— a breakdown of energy costs by fuel type and end user to enable potential savings in specific areas to be estimated. Effort can then be concentrated where the best opportunities are anticipated.

— recommendations for future opportunities, perhaps requiring major investment or additional study. It may be possible to set a programme for the introduction of new technology or to adopt the best current practices for controlling energy use.

— suggested operational changes that might result in improved plant reliability or availability. Benefits could arise from reduced maintenance or increased productivity.

— estimates of projected energy consumption needed when setting budgets for fuel purchase, or estimating the cost of providing a specific service.

— long-term options involving major refurbishment or influencing future policy on design and operation. A strategy may also be developed with the flexibility to cope with changes in the building use or choice of fuel type.

— a plan for developing energy management, including staff and budget requirements.

3.2.2 Working conditions

Some measures can be implemented in a way that improves the quality of a service, but does not necessarily reduce the energy costs:

— Comfort might be improved by draughtproofing, insulating the building fabric, resetting controls, providing additional controls or installing alternative systems. Changes in the temperature, humidity or lighting levels may be desirable. Productivity can be increased where the occupants are more satisfied with the working environment.

— Closer control of space conditions can be essential to the effective operation of buildings or equipment and could result in higher standards of quality and safety.

3.3 Environmental benefits

The main environmental benefits of using energy more efficiently are:

— reduction of CO_2 and other emissions that can be harmful to the environment

— reduction of national energy demand

— conservation of natural resources, particularly fossil fuels.

Every energy user has an effect on the environment. All can contribute to and share in environmental benefits.

The potential long-term consequences of irreversible environmental damage and exhaustion of fuel supplies demand action now. Attention to this matter is likely to become increasingly focused as the commercial incentives to improve efficiency and become 'environmentally friendly' grow.

3.4 Summary of section 3

* An audit and survey can have financial, operational and environmental benefits.

* Financial benefits can arise from simple no-cost measures and investment in energy efficiency. Measures are appropriately categorised into 'no/low-cost', 'medium-cost', and 'high-cost'.

* Operational benefits, such as the increased supply of management information and improved working practices, can improve productivity and lead to further financial benefits.

* Every energy user has a responsibility to prevent environmental damage and to conserve natural resources.

4 Choosing a suitable approach

4.1 Auditing, surveying and monitoring

A preliminary audit will form the basis of the initial decisions on how to proceed and whether assistance will be needed. This first stage, using existing site records and information, will assist in deciding the scope of survey to be performed and the precise resources needed. A site survey will enable a more detailed breakdown of costs to be made, will identify the options for improvements and will indicate the requirements for monitoring. Once established, the data provided by monitoring can be used in future audits.

The precise strategy adopted will need to suit the particular type of site or sites under investigation, whether single buildings, multi-building sites, or multi-site organisations.

The detailed procedures involved in the processes of auditing and surveying apply almost equally to a given type of building, whether it stands by itself or is part of a group.

4.2 Single buildings

If the single building has utility metering then its energy consumption can be determined from invoices, and the total annual energy costs calculated. The major types of plant and systems installed, e.g. boilers, chillers, must then be identified to assess the requirements for skills and resources. It will also be necessary to consider what expenditure is justifiable for the survey and to establish specific tasks and objectives. Recognition of the principal uses of energy — typically space heating, air conditioning, lighting, hot water services, mechanical ventilation, catering and where appropriate, process energy — will assist in deciding the priorities. The survey should normally deal with the building as a whole, so that all energy use can be accounted for, but placing emphasis on the principal uses of energy, particularly those expected to offer the best opportunities for savings.

4.3 Multi-building sites

While it should be possible to identify the total site energy consumption and cost quite easily from invoices, a typical multi-building site does not have utility metering for each building. An immediate allocation of site energy to each building might be possible if reliable information is available from sub-metering. The preliminary audit should at least identify the major energy-using buildings and the principal services on which a survey will be concentrated. All buildings should be accounted for at this stage, even if only to identify them for inclusion or exclusion in subsequent detailed audits and surveys.

The strategy for surveying will be decided with reference to the site layout and arrangements of the services. Minor buildings can often be excluded from further investigation. They may, however, be connected to a site distribution system which needs review. It would then be necessary to specify not only which buildings but also which site distribution services are to be surveyed. Effort should still be concentrated on the principal buildings or systems, but taking account of possible interactions.

A major site does not necessarily demand correspondingly substantial resources. An approach that examines individual buildings or groups of buildings progressively can be performed if limited manpower and test instrumentation are available. In any event, the size of the survey team should be such as to cause minimum disruption of the site. With proper co-ordination each building can be given appropriate consideration as part of a complete site.

4.4 Multi-site organisations

The multi-site organisation is more than just a collection of single and multi-building sites. Preparation of an overall group-wide energy strategy is needed to ensure the most effective use of time and resources. A preliminary audit will identify the basic energy costs at each site. For an organisation with a large number of small sites the volume of available data and the problems of missing invoices can make this an exercise in itself. The reliability of the data should always be confirmed before commitment to a programme of surveys. It is not unknown for information to be supplied for the wrong building. Information on building area can often be out-of-date or unreliable.

Priority should be given to sites with the greatest potential for reducing energy costs. A site with a large energy bill and poor apparent energy performance, as revealed by the preliminary audit, is expected to offer the best opportunities. When no indication of performance is available then priorities may have to be decided largely on the basis of energy cost. Generally a small percentage saving at the largest site will give the maximum value of savings but smaller savings at sites with a poor energy performance may be cheaper and easier to achieve.

It should be recognised that arbitrary measures of performance are useful for selecting sites likely to offer the best potential savings. However, they cannot take into account all the circumstances that affect comparison between any two sites. Even sites with a good apparent performance can still present excellent opportunities for savings. A degree of judgement must always be applied in deciding the best survey programme.

Many of the detailed audit and survey procedures will be essentially the same as for individual buildings or sites, but the priority for implementation could well differ.

The strategy may be determined by the availability of resources to perform surveys, the allocation of capital to implement recommendations and general matters of policy. In a very large organisation, a phased programme covering two or three years is likely to be needed to investigate the whole building stock. Pilot projects can help to determine appropriate standards and upgrading policies and ensure that the best approach is adopted.

Recommendations for individual buildings or sites should also take into account the wider implications for the organisation. It can be advantageous to combine the implementation of similar works on different sites into a single contract. A standard policy on design or control arrangement could also affect the preferred options for each location.

4.5 Resources

The requirements for carrying out surveys can be summarised as:

— staff with relevant knowledge/skills

— time allocated to perform the tasks involved

— technical equipment for any measurements

— finance for assistance in the above areas and to implement recommendations

— technical and operational information on buildings/ plant/ services.

4.5.1 Knowledge/skills

For most buildings, extensive technical expertise is not essential for the work involved in collecting audit data or for making basic survey observations. Simple recommendations, particularly for good-housekeeping measures, can almost always be made. Technically based recommendations may also be within the scope of someone with a sound understanding of the systems under investigation and the ability to interpret data. Those without relevant experience should, however, proceed with caution.

The most competent person to perform an energy audit and survey is a professionally qualified engineer experienced in this field. Large organisations without such staff may find it advantageous to have someone trained or even to make a new appointment in this role. Those with an annual energy bill of £1 million will usually find it cost-effective to employ a full-time equivalent member of staff. One advantage of an in-house capability is that it allows organisations to build a knowledge base about their buildings. Work performed in-house should also be cheaper than that contracted out. Financial or technical skills within an organisation will prove particularly useful given guidance and co-ordination. In-house staff can apply a particular knowledge of their organisation and its methods, although care is needed to avoid a restricted outlook.

For organisations where suitable staff are not available it will be necessary to commission specialist assistance. The use of an outside specialist can have several benefits, not the least being a shorter timescale. A fresh approach, free of internal constraints, will often result in new opportunities. An independent, expert view can also provide additional credibility in presenting the case for investment. Outside assistance is most commonly obtained from consultants. An outline of the types of services that are available is given in Appendix A2.

4.5.2 Timescale

The time taken to perform an audit and survey depends on the availability of energy data, the size of the site, and the complexity of the systems. An audit might take only a few hours for a simple site for which information is readily available. On more complex sites, a week or more could easily be spent simply on analysing invoices and records.

There are no firm guidelines for deciding the time that should be spent on a site survey — it will reflect the complexity of a site, the availability of resources and the expense that can be justified. An estimate can be built up by considering the individual elements to be examined. The largest of sites could need the equivalent of one man–year to survey comprehensively, or preferably a small team to achieve a shorter survey period. A concise survey of a small building could be completed by one person within a day.

Time must be allowed both for those conducting a survey and those contributing in other ways. This could range from providing information to simply acting as an escort. Even with outside assistance some in-house staff participation will always be needed. The greater the co-operation, the better the survey. Staff should therefore be encouraged to make a positive contribution.

4.5.3 Metering and instrumentation

Measurement provides the foundation for understanding the flow of energy on site. The use of metering and instrumentation will enable a quantitative analysis to be made of energy use and of the quality of service maintained. With skilled application and experience in the interpretation of results far more knowledge is likely to be gained than from observation alone. Attention is drawn to the importance of using correctly calibrated instruments to obtain reliable information.

Temporary test equipment can readily be purchased or hired for most purposes where a definite need for accurate, measured data is identified. It should be used only to the extent really necessary to draw sound conclusions. A correctly performed test will avoid the production of excess data from too many readings or too long a recording period.

The measurements most commonly required are as follows.

Environmental conditions

— room temperature †

— relative humidity †

— lighting level

Electricity

— load (kW/kVA) †

— current

— hours run

Air handling

— flow rate (grilles and ducts)

— supply temperature

— relative humidity

— differential pressure

Piped services

— steam †/gas †/oil †/water † flow rate

Boiler plant (flue gas analysis)

— temperature

— O_2/CO_2 content

— CO content

— smoke number

Selective spot readings are always useful, but parameters marked with a dagger often need to be recorded over a period to discern patterns.

A concise survey may require minimal instrumentation. Comprehensive surveys are normally expected to include measurement of principal energy flows and performance assessment of major plant. Reliable measurements of building areas and volumes are also needed for detailed assessments.

4.5.4 Finance

The overall costs of investigating energy use can be divided into three distinct options, each of which may be funded separately.

— Audit and survey investigations

— Implementation of recommendations

— Monitoring procedures.

Audit and survey costs

The cost of in-house work depends on internal accounting procedures. For outside assistance the cost will typically be a fixed sum or a function of a shared-savings scheme. In assessing value for money for these services, similar criteria should be adopted to those applied to the rest of the organisation's business.

Whichever route is followed the total audit and survey costs are not expected to exceed a small percentage of the annual cost of energy under investigation. A figure between 3% and 5% is not uncommon, perhaps falling towards 1% for energy-intensive sites with multi-million pound energy bills. More than 10% is unlikely to be justified even for small sites where the energy bill may be less than £10 000. Note that the percentage may remain at the higher level for an organisation with many small sites, even though the total energy bill may be several million pounds. The precise figure will depend largely on the depth of study undertaken. If audit and survey recommendations are implemented fully, the costs of these services are normally recovered from savings in the first year. The survey may also highlight measures requiring further study or design. These further study or design costs should also be identified at the time of implementation.

Funding recommendations

Recommendations fall into the categories of implementation cost discussed in section 3.1, ranging from no-cost to high-cost. Guidelines should be agreed at the start of a survey to define these categories. Remember that measures with no capital cost can still require some management time. Constraints which will affect the options that can be put forward should also be indicated. These might include:

— availability of funding; internal or other resources; contract energy management option

— categories of cost requiring executive approval or suitable for immediate authorisation

— investment return criteria; simple payback related to life expectancy of measure; discounted cash flow

— cost of in-house or outside assistance when further design or study is needed.

Having established which options meet the criteria, they should then be arranged in order of priority for implementation. A programme of expenditure can then be agreed, phased over several years if necessary for major works, and budgets planned accordingly.

The criteria should equate with those applied to return on investment in the rest of the organisation's business, rather than some arbitrary fixed payback period. Rejected schemes should still be noted for future review. They may become worthwhile if circumstances change.

Monitoring procedures

Monitoring is an essential element of energy management. It usually requires an initial investment in additional metering, followed by a regular commitment to the collection and analysis of data. Use of a microcomputer is recommended to keep the time spent in routine analysis to a minimum.

If correctly interpreted, the results of monitoring will provide a measure of the savings actually achieved. The information will also help to ensure that savings are maintained in the long term. By highlighting opportunities for further savings the value of monitoring can be increased enormously.

Monitoring costs should not exceed a small percentage of the annual energy cost. They must be balanced against the value of savings and information obtained.

The cost of installing metering is a 'one-off'. It should not normally exceed a small percentage of the annual energy cost. Although metering in itself produces no savings, in conjunction with monitoring and target setting it can be a very cost-effective way of identifying savings and opportunities. Investment in electricity meters is usually more profitable than investment in, say, heat meters which tend to be more expensive and less reliable.

4.5.5 Technical and operational information

Reliable site information can assist considerably in making the best use of the time spent on an audit or survey. It should always be made available to those conducting the study, as should the personal knowledge of the site staff of particular systems or operating practices. Examples of the possible types of information are:

— drawings showing building or system layouts, dimensions or construction details

— schedules of plant or equipment with ratings or performance figures

— operating or maintenance manuals

— knowledge of occupancy patterns and working practices, particularly where the survey period cannot cover seasonal changes or maintain continuous observation

— floor space statistics; often available where there are tenancy agreements or cleaning contracts but often unreliable.

4.6 Summary of section 4

* Auditing, surveying and monitoring are linked as components of an overall strategy for effective energy management.
* An approach must be developed to suit each individual site or organisation under investigation — it is particularly important to decide priorities where multiple sites or buildings are involved.
* Preliminary work can be undertaken in-house by someone without technical expertise and, given guidance, developed by those with relevant skills.
* Large organisations without suitable staff available may find it worthwhile to train or appoint new staff.
* Specialist professional assistance can complement in-house resources and provide a fresh, independent viewpoint.
* The time taken to conduct an audit and survey may range from less than one day for a small building, to several months for a comprehensive study of a major site.
* Even when using outside assistance some time must always be allowed for the participation of in-house staff.
* Additional metering and test instrumentation are likely to be needed for detailed surveys.
* A complete project requires funding at three stages — initially for the audit and survey, then to implement the recommendations and finally for continued monitoring.
* Audit and survey costs of between 3% and 5% of the annual energy cost are typical, but the percentage may be higher for small buildings and lower for large sites.
* Financial constraints and investment criteria for energy saving measures should be established at the outset.
* Monitoring schemes require some investment to set up, a small cost in manpower to maintain and the delegation of individual responsibilities to achieve results.
* Monitoring provides essential management information on progress when savings measures are implemented and can help to identify further opportunities.
* The availability of reliable information on a building or site can help to ensure that survey time is spent most effectively.

5 Preliminary audits

A preliminary audit seeks to establish the quantity and cost of each form of energy used in a building or site. If reliable data exist, a breakdown of energy use by area or by service may also be possible. Collated information should be used to prepare a plan for further action leading to a site survey. An audit period of one year is recommended, normally corresponding to the most recent financial year or other convenient period.

The main steps can be summarised as:

(*a*) collection of data

(*b*) analysis of data

(*c*) presentation of data

(*d*) establishing priorities.

5.1 Data collection

Invoices are the principal source of data, supplemented where possible by site records.

5.1.1 Invoice data

Invoices are an essential source of information on energy input, costs and tariffs. Original invoices should always be obtained in preference to transcribed extracts as important data or errors can otherwise be missed. Actual meter readings and dates will assist with checking and interpretation.

— Collect copies of all monthly and quarterly invoices for energy used during the full audit year, not just those for which payment was made in this period.

— For fuels supplied in bulk — oil, solid fuel, liquefied petroleum gas, etc. — collect all invoices or delivery notes relevant to use during the audit period. Deliveries before the start of the period may have to be included but a delivery at the very end can usually be excluded.

— Check that all metering or supply points can be identified from invoices and that all supplies can be accounted for.

— Ensure that data for each type of energy refer, as closely as possible, to the same period.

— Note any estimated readings. Additional invoices with real readings should be collected for comparison if there are more than one or two estimates in the audit period.

— Approach suppliers for assistance if invoice information is inadequate or unavailable.

5.1.2 Electricity

Small commercial premises are normally supplied under a quarterly billed non-domestic tariff. Most larger commercial or industrial premises are supplied under monthly billed demand tariffs. An invoice for a small site may show little more than the basic unit cost and a quarterly standing charge.
Sites which are invoiced monthly generally receive additional details which vary between electricity suppliers.

METER READINGS PRESENT	METER READINGS PREVIOUS	ADVANCE	CONSTANT	CONSTANT DEC	UNITS CONSUMED
252072	204740	47332	15	000	709980
192759	175636	17123	15	000	256845
			TOTAL		966825

DATE READ

DESCRIPTION OF CHARGE	NUMBER OF UNITS, KVA OR KW	RATE OF CHARGE £		VAT	CODE	AMOUNT OF CHARGE £	p
UNIT CHARGE	709980	0	03860	0		27,405	23
FUEL VARIATION	709980	0	0015675CR	0		1,112	89CR
MAXIMUM DEMAND STEP 1	300	1	74	0		522	00
MAXIMUM DEMAND STEP 2	1665	1	66	0		2,763	90
SUPPLY CAPACITY	300	0	62	0		186	00
SUPPLY CAPACITY STEP 2	1700	0	48	0		816	00
UNIT CHARGE	256845	0	01850	0		4,751	63
FUEL VARIATION	256845	0	0015675CR	0		402	60CR

YOUR REFERENCE NUMBER	FUEL COST PER TONNE	MAXIMUM DEMAND THIS MONTH	TOTAL CHARGE NOW DUE FOR THE MONTH OF FEB
	4725 p	1965	34,929 27

Figure 5.1 Monthly electricity invoice

The example in Figure 5.1 includes many of the features that might be found. The information included is generally as follows:

— Date of meter reading. Special note is needed of readings that are estimated, but these are only expected to occur with quarterly accounts.

— Present and previous meter readings with advances. There will be two sets of meter readings where a day/night tariff is in force, as in Figure 5.1.

— If applicable, a constant for multiplication of the meter advances to give the unit consumption in kWh. This is sometimes indicated only on the invoice, not on the meter itself. In Figure 5.1 there is a constant of 15.

— The charge(s) for each unit consumed which will be different for the day and night (3.86p and 1.85p respectively in Figure 5.1), and sometimes evening and weekend units if they are metered separately. Blocks of units may also be charged at different rates. Quarterly billed non-domestic tariffs usually have a higher charge for a primary block of typically 1000 units per quarter. Maximum demand tariffs may have a higher charge for blocks of up to 300 kWh per kVA or kW of maximum demand each month.

— The fuel variation charge, if applicable, will vary the cost of all units depending on a declared cost of generating station fuel. This is the present method of making fine adjustments to published tariffs. The variation charge in Figure 5.1 is a credit of 0.1565p/kWh.

— For tariffs related to load, the maximum demand occurring during the month in kW or kVA. For maximum demand charges, this may be divided into steps with different costs. These charges are applied to penalise users who place heavy demands on the grid during peak periods. Figure 5.1 shows a maximum demand of 1965 kVA, the first 300 kVA being charged at £1.74/kVA, the remainder at £1.66/kVA.

— The supply capacity, for which an availability charge is made by most supply companies, indicates the nominal limit of maximum demand. It must be adequate to meet the maximum power requirement but sometimes can be adjusted, in steps, to minimise surplus chargeable supply capacity.

— The total cost of all items which, when related to the total number of units consumed, gives the overall cost per unit.

Leaflets providing details of the applicable tariff structures should be obtained from the relevant supplier. Not only do these assist in understanding the invoice information but will be essential when assessing the potential savings from subsequent actions.

5.1.3 Natural gas

Premises consuming in excess of 25 000 therms per annum are normally invoiced monthly. Prices may be fixed or index linked for the contract period and depend on the volume of consumption and the number of premises supplied. Major sites may have a combination of firm and interruptible supplies, invoiced separately, with prices that differ widely. In some cases consumptions may be aggregated under a single contract. Firm supplies are subject to seasonal pricing factors. Small consumers are normally invoiced quarterly. Generally, a larger user can expect to pay a lower price per therm but a higher monthly standing charge than a small consumer.

Figure 5.2 shows a typical example of a monthly gas invoice. The layout of invoices varies between suppliers, but the information included is generally as follows:

— Date of meter reading or estimate. Special note is needed of readings that are estimated.

— Present and previous readings with advances, usually in hundreds of cubic feet.

— A multiplier to convert hundreds of cubic feet to heat units (therms). This multiplier is derived from the declared calorific value in Btu per standard cubic foot which is also included on the invoice. Thus if the declared calorific value were 1030 Btu/ft³ then the multiplier would be 1.03 to convert hundreds of cubic feet to therms. As the density of the gas depends on its temperature and pressure, the measured volumes on major supplies are corrected, usually automatically at the meter, to standard conditions of 15.5°C and 1013 mbar. Where the readings are not automatically cor-

ACCOUNT REFERENCE NUMBER	DATE OF ACCOUNT
	20/06/ (TAX POINT)

'E' - ESTIMATED READING
'C' - CUSTOMERS OWN READING
If this account is based on an estimated reading which is substantially higher or lower than the true reading of the meter you should indicate the exact reading in the appropriate space below and return the account immediately for amendment.
Any underestimation of consumption in this account will otherwise result in a correspondingly higher consumption being charged when the next factual reading is taken.

READING OR INVOICE DATE	METER READING: PRESENT	METER READING: PREVIOUS	GAS SUPPLIED: CUBIC FEET (HUNDREDS)	GAS SUPPLIED: THERMS	PRICE PER THERM	AMOUNT	VAT CODE	VAT CHARGES
18 6	8092	5783	2309	2345.944	37.000	868.00	A	0.00
	STANDING CHARGE					9.90	A	0.00
	TOTAL VAT CODE A @ ZERO%					0.00		
	TOTAL VAT							0.00

CALORIFIC VALUE		TARIFF	
37.9 MJ/m³	1016 Btu/ft³	NON-DOMESTIC	£ 877.90 AMOUNT NOW DUE

Figure 5.2 Monthly gas invoice

rected, a separate pressure correction factor for the particular meter and location is sometimes stated.

- A scheduled price per therm.
- A fixed monthly or quarterly charge may be included for consumers using up to 25 000 therms per annum. The additional monthly charge for larger users may depend on the annual consumption and the choice of a firm or interruptible supply.
- The total cost.

5.1.4 Liquefied petroleum gases (LPG)

LPG invoices show the deliveries on a mass basis; the calorific value may be obtained from the supplier.

5.1.5 Steam

Imported steam is usually charged by mass, although it will almost certainly be measured by a meter working on a volume basis. The meter output data therefore contain a conversion from volume to weight via a known density which should be determined from actual steam conditions.

Problems of accuracy and correct calibration are very common, particularly under widely varying or fluctuating load conditions, or if steam pressure has been changed since the original installation.

Particular attention must therefore be given to any correction factors applied when examining records of steam consumption.

5.1.6 Fuel oils

Delivery invoices contain:

- the delivered volume in litres
- the volume corrected to the standard condition of 15.5°C
- the cost per standard litre
- the total cost.

The calorific value can be obtained from the supplier.

5.1.7 Solid fuel

Short-term determinations of solid fuel consumption are generally avoided due to errors in assessing coal stocks. Coal delivery notes quote weight and cost, but generally do not include data for calorific value. However, the data can be obtained either from the supplier or direct from British Coal. Variations in water content change calorific value proportionally and average values will therefore have to be assumed.

5.1.8 Heat

Where heat is imported this is usually in the form of medium- or high-temperature hot water. Heat meters monitor the volume flow rate and the temperatures of water entering and leaving the premises. By integrating flow rate with temperature difference, to give heat flow rate, a meter can provide an indication of heat energy consumption, typically in kWh.

Changes from design conditions can affect accuracy, particularly very low flow rates or small temperature differences which make metering errors more significant.

5.1.9 Site energy records

For large sites or buildings, site energy records may provide details of any sub-metered energy consumption, stock levels or use of bulk fuels. In small buildings they may be able to supplement incomplete or estimated invoice data.

- Collect any summaries of invoice data which have already been produced for the audit year and, to show overall trends, for the previous two years. Summaries may need validation before they can be used for analysis.
- Collect records of energy for the audit year. Monthly or weekly summaries are more useful than daily log books as anomalies may arise in the shorter term and such detailed analysis is not expected for the audit.
- Confirm the arrangement of all sub-meters relative to their respective main meters, including any sub-meters for which records are not available but from which additional data could be obtained. Use of schematic diagrams of all metering systems is recommended as in Figure 5.3.

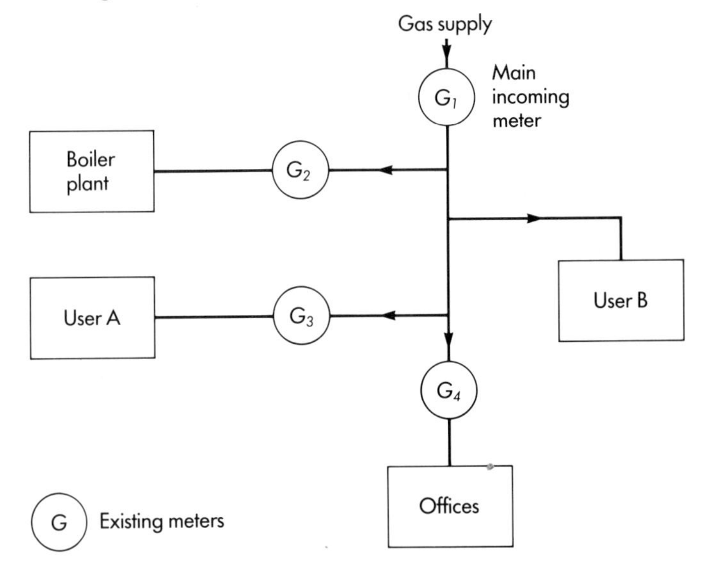

Figure 5.3 Schematic diagram of meter layout. (Consumption of User B can be determined by difference $G_1 - (G_2 + G_3 + G_4)$.)

- Consider whether correction or multiplication factors should be applied and check the units of energy measurement. In particular note the addition of noughts following meter readings. The appropriate correction may be small enough in many cases to be unnecessary but this has to be established. Sub-meter readings in particular are likely to need correction, especially for pressures changes in gas or steam distribution. Further information on correction factors is given in Appendix A8.
- Where reliable sub-meter records are available, try to determine the consumption of users without sub-meters

by difference as in Figure 5.3. Care is needed as cumulative errors can often give inaccurate results. Be especially wary of the accuracy of old or neglected sub-meters.

5.2 Data analysis

5.2.1 Annual energy input

Analysis of consumption and cost from invoices assists the selection of priorities for survey. Data for individual sites should be analysed separately as follows:

- Identify each individual energy type to be analysed.
- Distinguish between interruptible and non-interruptible gas supplies which are likely to show a significant difference in price and use.
- Make special note of estimated readings, which can occur commonly for small electricity and gas supplies. Estimates are often based on the consumption for the corresponding period in the previous year and could well be inaccurate.
- Ensure that recorded annual totals apply to a full twelve-month period. An estimate in the middle of the audit year will not affect the recorded annual total. In the first or last period, an estimated or mistimed reading will need some interpretation. Interpolation can often be applied successfully to monthly readings, simply by assuming a uniform rate of consumption between the dates of two known readings. For quarterly readings, the results of this method are far less reliable and further analysis may be restricted.
- Convert the consumption of each energy type to a common unit (GJ) using the standard conversion factors in Table 5.1. The typical calorific values may be assumed if actual values are unavailable.
- Calculate the percentage breakdown of total energy consumption and cost by energy type and determine the average overall cost per gigajoule of each energy type to indicate its relative significance.
- Prepare a table on the model of Table 5.2 showing the total annual consumption and cost of each fuel type for the audit year. It is useful, for reference, to include the respective proportions of total cost and energy and the overall unit costs for comparison.
- Prepare pie charts of the type shown in Figure 5.4. Illustrate the energy and cost contributions of each energy type.
- Where the previous year's energy data are available, a comparison with the audit year should be made, as this may indicate any major overall trends as illustrated in Table 5.3.

Table 5.1 Calorific values and factors to convert to standard units

Fuel type	Typical calorific value (gross basis)	Conversion factors		
		Multiply by		
		From	Factor	To
Electricity	1 kWh/unit	kWh	3.6×10^{-3}	GJ
Natural gas	1.01 therms/100 ft^3	therms	1.055×10^{-1}	GJ
Gas oil (Class D)	38 MJ/litre	MJ	1×10^{-3}	GJ
Heavy fuel oil (Class G)	42 MJ/litre	MJ	1×10^{-3}	GJ
Coal	27–30 GJ/tonne	m^3	8.2×10^{-1}	tonnes
Propane	50.0 GJ/tonne	litre†	5.0×10^{-4}	tonnes
Butane	49.3 GJ/tonne	litre†	5.7×10^{-4}	tonnes

†Liquid at standard condition

Table 5.3 Changes in annual energy use

Year	Consumption (GJ)	Change (%)
1987/88	5000	—
1988/89	4800	−4.7
1989/90	5100	+6.2

Note that at this stage no correction has been made for variable factors which might affect the comparison.

5.2.2 Assessment of performance using yardsticks for annual energy consumption

The total annual energy use of a building can be used to calculate a measure of energy performance known as the 'Normalised Performance Indicator' (NPI).

Comparison of the NPI with yardsticks for buildings of similar type can then indicate whether there is likely to be a good opportunity for improvement.

Yardsticks for energy consumption should be recognised as providing useful guidance in setting priorities but should not be taken as absolute values to be achieved. The actual potential for improving performance can only be determined by on-site investigation.

The series of booklets *Energy Efficiency in Buildings*[1] describes energy performance in terms of three categories (Table 5.4).

Table 5.2 Table of annual energy input for audit year 1988/9

Energy type	Purchased units	Consumption			Cost	
		(GJ)	(%)	(£)	(%)	(£ GJ^{-1})
Electricity	902 500 kWh	3 249	14	37 400	57	11.51
Gas-firm	5 500 therms	580	3	1 880	3	3.24
Oil-class G	440 000 litres	18 480	83	26 200	40	1.42
Totals	—	22 309	100	65 480	100	2.94† av.

†Average cost (£/GJ) is defined as total cost divided by total energy purchased.

Table 5.4 Building energy performance classification[1]

Category	Attributes
Good	Generally good controls and energy management procedures although further energy savings are often possible.
Fair	Reasonable controls and energy management procedures, but significant energy savings should be achievable.
Poor	Energy consumption is unnecessarily high and urgent action should be taken to remedy the situation. Substantial energy savings should result from the introduction of energy efficiency measures. There may be valid reasons due to special circumstances.

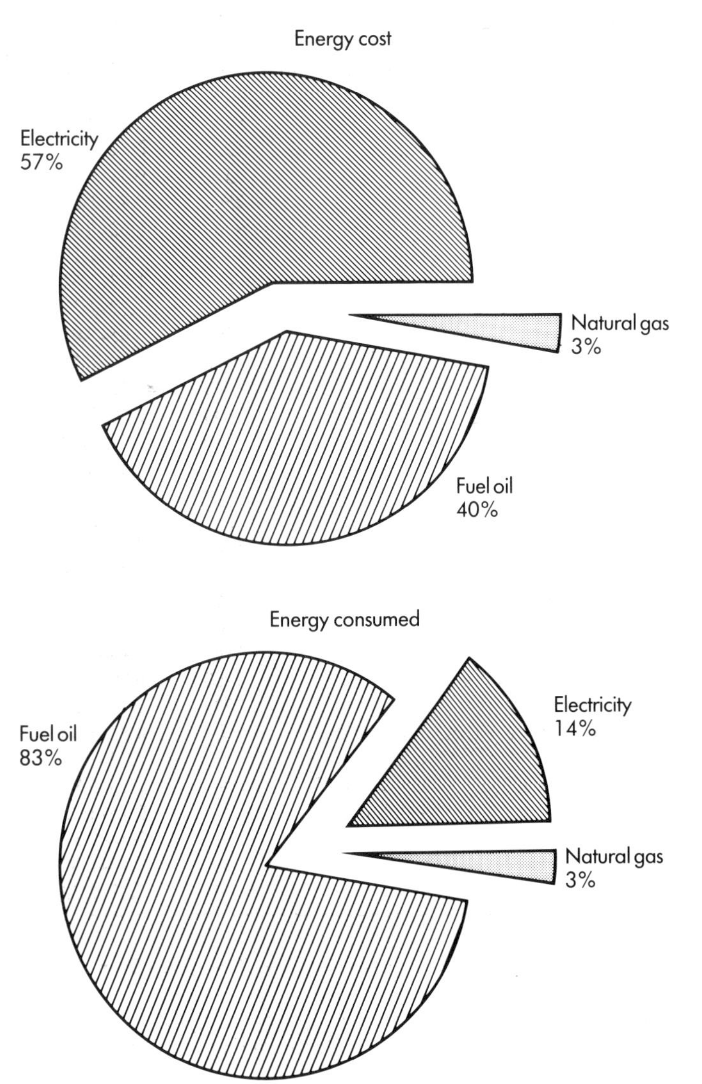

Figure 5.4 Pie charts of annual energy cost and consumption

A category may be assigned to most buildings for which the total energy use and other basic data can be established. The full procedures are described in the *Energy Efficiency in Buildings* booklets[1]. Readers are advised to consult these booklets but the steps are summarised here for convenience.

1 Determine total building energy use in common units as described in section 5.2.1. Note that the gigajoule is the standard unit for energy use in this *Manual.*

2 Determine the energy use for space heating. If it is not metered separately assume a percentage of the fossil fuel energy as given in Table 5.5. Multiply the energy use figure by the Weather and Exposure factors given in Table 5.6 to give the corrected energy use for space heating. (See Appendix A7 for information on degree–days).

3 Add the non-heating energy use to the corrected space heating energy use. Multiply the total by the Hours of use (or Shift occupancy for factories) factor given in Table 5.6 to give the normalised annual energy consumption.

4 Divide the normalised annual energy consumption by the building floor area in m^2 to give the NPI. Floor area should exclude completely untreated areas and is normally less than the gross area included within the building envelope.

5 Compare the NPI with the yardsticks given in Table 5.8 and classify the performance.

A worked example of the calculation of a normalised performance indicator for a primary school is given in Table 5.7.

Table 5.5 Proportion of fuel used for space heating and hot water assumed to be attributable to space heating when calculating performance indicators[1]

Building type	Proportion of fuel used for space heating and hot water attributable to space heating (%)
School	75
Hospital, nursing home	50
Other health care	75
Further/higher education	75
Office	75
Sports centre, no pool	75
Sports centre, with pool	65
Swimming pool	55
Library, museum, gallery	70†
Church	90†
Hotel	60
Bank, agency	75
Entertainment	75
Prison	60
Court, depot, emergency services building	75
Factory	80

†Percentage of total energy consumed within the building.
Percentages may also be appled to all-electric buildings.

Table 5.6 Correction factors to be appled to actual energy use to normalise the calculated performance indicator[1]

Parameter	Correction	Factor
Weather†‡	$\frac{\text{Standard degree–days (2462)}}{\text{Degree–days for energy data year}}$	
Exposure†	Sheltered — city centre	1.1
	Normal — urban/rural	1.0
	Exposed — coastal/hilly site	0.9
Hours of use	$\frac{\text{Standard hours of use (Table 5.8)}}{\text{Actual annual hours of use}}$	
Factory occupancy	Single shift, 5-day week	1.00
	Single shift, 7-day week	0.83
	Double shift, 5-day week	0.77
	Double shift, 7-day week	0.72
	Continuous working	0.72

†Applies to space heating energy only.
‡Alternative regional weather factors are sometimes quoted for shops.
No corrections need be applied to catering establishments.

Table 5.7 Example of calculation of normalised performance indicator

Data	Building type	Primary school, no indoor pool	
	Location	Urban	
	Hours of use	1500	
	Energy intake	16000 therms natural gas	
		14000 kWh electricity	
	Floor area	2400 m^2	
	Degree–days for data year	2450	
1	Total energy use	$16000 \times 0.1055 =$	1688
		$14000 \times 0.0036 =$	50
			1738 GJ
2	Space heating energy	$1600 \times 0.75 =$	1266 GJ
	Weather correction	$1266 \times \frac{2462}{2450} =$	1272 GJ
	Exposure correction	$1272 \times 1.0 =$	1272 GJ
3	Non-heating energy	$1738 - 1266 =$	472
	Add corrected space heating energy		1272
			1744 GJ
	Hours of use correction	$1744 \times \frac{1480}{1500} =$	1721 GJ
4	Calculate NPI	$\frac{1721}{2400} =$	0.72 GJ/m^2
5	Assess performance		FAIR

Table 5.8 Yardsticks (GJ/m^2 per year) for annual energy consumption of common building types[(1)]

Building type	Standard hours of use per year	FAIR performance range
Nursery	2290	1.33–1.55
Primary school, no pool	1400	0.65–0.86
with pool	1480	0.83–1.12
Secondary school, no pool	1660	0.68–0.86
with pool	2000	0.90–1.12
with sports centre	3690	0.90–1.01
Special school, non-residential	1570	0.90–1.22
residential	8760	1.37–1.80
Restaurants	—	1.48–1.87
Public houses	—	1.22–1.69
Fast-food outlets	—	5.22–6.30
Motorway service areas	—	3.17–4.32
Department/chain store (mechanically ventilated)	—	1.87–2.23
† Other non-food shops	—	1.01–1.15
† Superstore/hypermarket (mechanically ventilated)	—	2.59–2.99
† Supermarket, no bakery (mechanically ventilated)	—	3.85–4.57
† Supermarket, with bakery (mechanically ventilated)	—	4.07–4.86
† Small food shop — general	—	1.84–2.09
— fruit & veg.	—	1.44–1.62
‡ Hospital, acute up to 20 000 m^2	—	—
acute over 20 000 m^2	—	—
long stay	—	—
‡ Nursing/residential care homes	—	—
‡ Dentists' surgeries	—	—
‡ Doctors' surgeries	—	—
‡ Clinics/health centres	—	—
University	4250	1.17–1.28
Polytechnic, residential	6000	0.83–1.13
teaching & admin.	3500	0.68–0.94
Colleges of further education	3200	0.83–1.01
Air conditioned offices, over 2000 m^2	2600	0.90–1.48
under 2000 m^2	2400	0.79–1.12
Computer centres	8760	1.22–1.73
Naturally ventilated offices, over 2000 m^2	2600	0.83–1.04
under 2000 m^2	2400	0.72–0.90
Swimming pool	4000	3.78–5.00
Sports centre with pool	5130	2.05–3.02
Sports centre/club, no pool	4910	0.72–1.22
Library	2540	0.72–1.01
Museum, art gallery	2540	0.79–1.12
Church	3000	0.32–0.61
Small hotels, guest houses	—	0.86–1.19
Medium-sized hotels	—	1.12–1.51
Large hotels	—	1.04–1.51
Banks, post offices	2200	0.65–0.86
Building societies	2250	0.47–0.61
Estate agents	2900	0.63–0.94
Insurance brokers	2300	0.50–0.76
Travel agents	2700	0.59–0.88
Cinemas	3080	2.34–2.81
Theatres (public)	1150	2.16–3.24
Bingo clubs	3500	2.27–2.77
Social clubs	3000	0.72–1.30
Prisons	8760	1.98–2.48
Police stations	8760	1.22–1.69
Fire stations	8760	1.58–2.23
Ambulance stations	8760	1.44–1.91
Crown and county courts	2400	0.79–1.08
Transport depots	2500	1.12–1.37
Factories — little process, under 2000 m^2	§	0.83–1.08
over 2000 m^2	§	0.94–1.33
Factories — heat gains, under 2000 m^2	§	0.68–0.97
over 2000 m^2	§	0.76–1.08
Warehouses, heated	§	0.54–0.97
Cold stores	§	1.80–2.43
Hangars	§	0.79–2.89

Notes

NPI below FAIR range is GOOD; NPI above FAIR range is POOR.

† Based on sales floor area.

‡ Figures for health care buildings are not currently available.

§ Factories data based on single shift, 5/6 day week, excluding process energy.

Yardsticks quoted in the EEO *Energy Efficiency in Buildings* booklets[(1)] for some common building types have been converted to GJ/m^2 per year and summarised in Table 5.8. Alternative values using kWh/m^2 per year are given in Appendix A6, along with some alternative yardsticks not based on floor area.

It must be stressed that the performance indicators are intended to allow comparison of energy performance between similar typical buildings. The POOR performers are most likely to offer the best cost effective opportunities, but improvement should even be possible for those classified as GOOD. In practice, the opportunities finally identified will not necessarily reflect these classifications.

As the average performance of buildings continues to improve, what is today considered GOOD may eventually be regarded as worse than average. Users should not therefore be content simply because a classification of GOOD is achieved under the present ratings, but should maintain efforts to improve efficiency.

5.2.3 Seasonal/monthly energy use

The pattern of energy use can be examined where monthly data are available either from invoices or site records. General trends or seasonal patterns could become evident, as in Figure 5.5, indicating priorities for further investigation. Records based on calendar months are preferred for this analysis as months derived from four- or five-week periods may be misleading. Weekly figures would introduce short-term fluctuation but might be useful for detailed analysis at a later stage. Quarterly figures can do little more than indicate the relative importance of summer and winter loads.

(*a*) Examine each metered consumption for which a full set of readings for the audit period is available. Decide whether the value of energy metered and the type of end use justify further analysis — an annual space heating bill of only a few hundred pounds may still offer worthwhile potential savings, while a major but essential electrical load might present little opportunity.

(*b*) Ensure that the readings for the selected meters cover consistent periods. With monthly figures, one or two days difference can be ignored for most purposes. Larger discrepancies can generally be corrected by simple interpolation or scaling. Care must be taken where the 'missing' or 'extra' days in a period are, for example, a weekend when a reduced allowance might be appropriate.

(*c*) For each selected data set plot a bar chart, of the type shown in Figure 5.5, of the consumption for each metered period. Try to account for any significant features:

— A seasonal or cyclical pattern could indicate major seasonal loads. This could influence the timing of a survey. Degree–days (see Appendix A7) can be plotted on the same chart, as in Figure 5.5, to show sensitivity to ambient temperature.

— General upward or downward trends can reflect changes in load or efficiency. They could also be attributed to changes in operating practice.

— The lack of a clear pattern where variations would normally be expected might suggest a lack of control. A range of effects could also be occurring at different times of year but may be indistinguishable.

— A steady base load can often be identified in boiler fuel consumption where the plant serves a mixed load. For example, in Figure 5.5 the area under the broken line represents the base load, perhaps due to domestic hot water and standing losses. The remaining consumption might then be attributed to space heating by deduction. In the example the base load is assessed at 18 GJ/month, suggesting a space heating consumption of $360 - (12 \times 18) = 144$ GJ/year. This technique should not be applied where the base load varies significantly from month to month.

— Obvious anomalies can often be explained by errors in the original data. Very high or low consumptions in particular months may, however, be due to excessive energy use or shutdowns respectively.

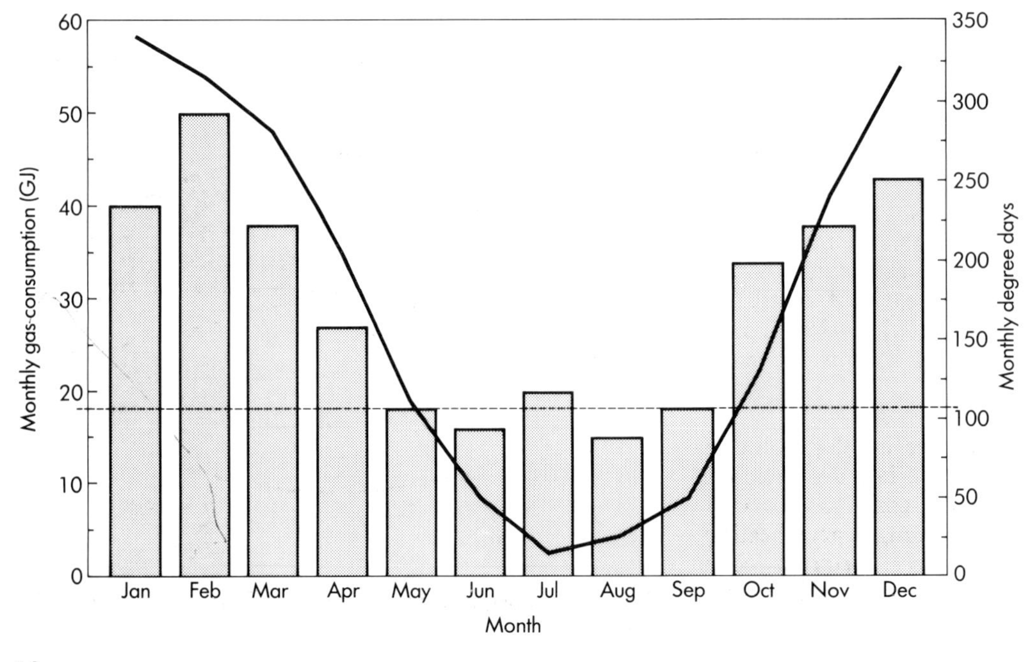

Figure 5.5 Bar chart of monthly fuel consumption with monthly degree–days shown as solid line (Annual total: 360 GJ)

(*d*) For consumptions including a significant element of space heating, plot a graph of monthly energy use against degree days (see Appendix A7) and try to fit a straight line to the points. The intercept, obtained as in Figure 5.6, can give an approximate indication of the base load, often attributable to standing losses or domestic hot water. Direct comparison with similar lines, using data from previous years plotted on the same graph, can provide a further indication of trends in consumption.

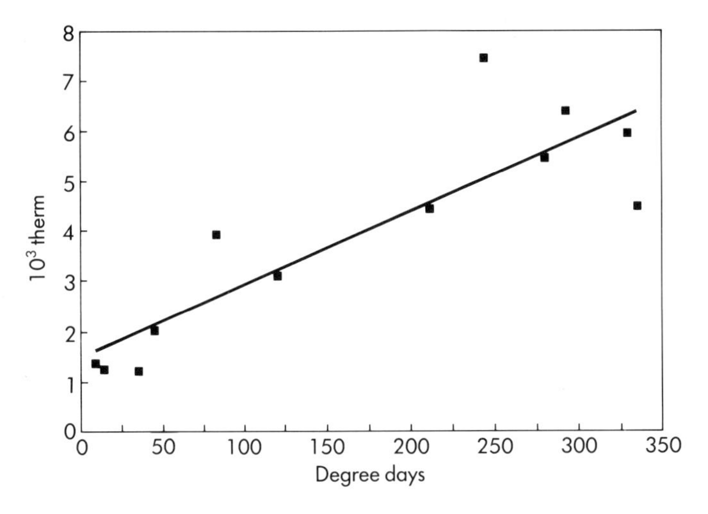

Figure 5.6 Graph of monthly energy use versus degree–days

The analysis using standard published degree–days is not strictly valid for intermittently heated buildings but can be accepted as a reasonable approximation at this stage. Wind speed and solar gain are two other influential factors which should be recognised but for which no allowance is made. System efficiency might also vary significantly with the heating load.

5.2.4 Breakdown of energy use

The first step in producing a breakdown of energy use is to establish a list of the main services or end users, as in the example of Table 5.9. A similar approach may be used whether examining different sections of a single building or different buildings on a multiple site.

Only the types of fuel used are noted initially, but this can be developed to include a quantitative assessment. Actual consumption figures can sometimes be given if an area or service has been sub-metered for the audit period. Otherwise, consumptions can be apportioned only provisionally without further site investigation.

Any consumption breakdown should be developed into a cost breakdown by applying the appropriate unit cost for the fuel types involved. The relative contributions may change substantially on a cost basis, although space heating will remain most significant in many cases. Electricity costs are likely to be dominant for heavily serviced buildings such as those with full air conditioning.

Table 5.9 Identifying energy uses

Energy use	Energy source	
	Electricity	Primary fuels
Space heating	✓	Gas and oil
Hot water service	✓ Summer	Gas
Air conditioning	✓	—
Lighting	✓	—
Process — heating	—	Gas
— power	✓	—
Other specified uses	—	—

A first estimate of the energy use for space heating might be obtained from Table 5.5 or by the base load analysis described in section 5.2.3. As an indication of the other likely major uses in a given type of building reference can be made to the typical examples(1) of energy use breakdown illustrated in Figures 5.7 to 5.29 together with the following notes. The examples are not intended to indicate what proportions of energy should be used in any building.

Schools

Figures 5.7 and 5.8 refer to a normal school building without catering or special sports facilities. An indoor swimming pool could account for 25% of the energy use in a school.

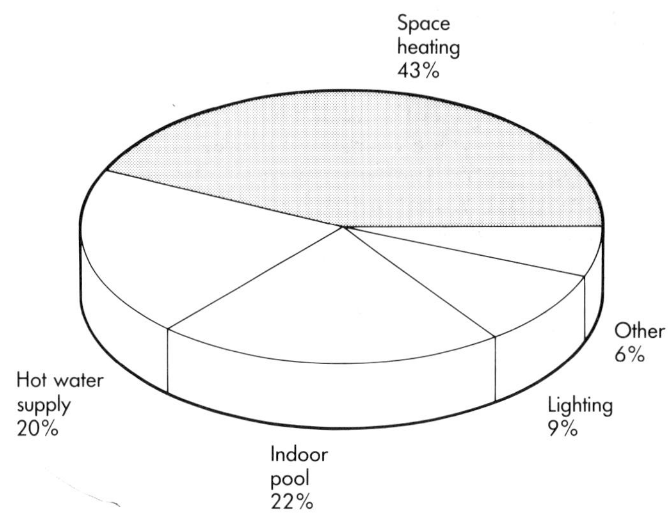

Figure 5.7 Energy use in a typical school with indoor swimming pool(1)

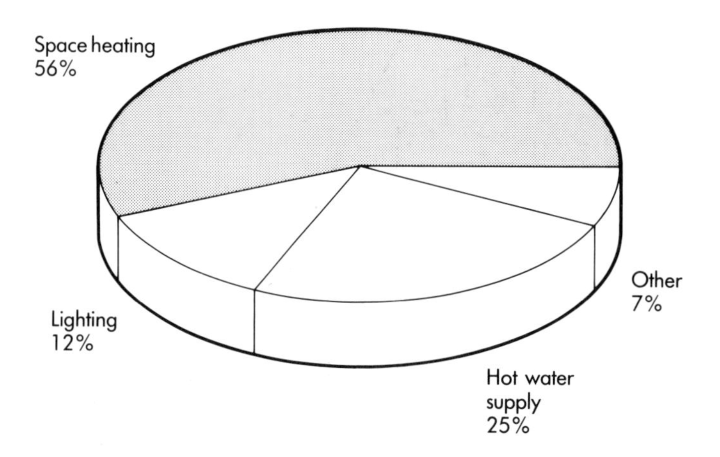

Figure 5.8 Energy use in a typical school without indoor swimming pool(1)

Catering establishments

Figures 5.9 to 5.11 show that space heating and lighting remain significant in many catering establishments but in a fast-food outlet, Figure 5.12, the energy use for cooking is dominant.

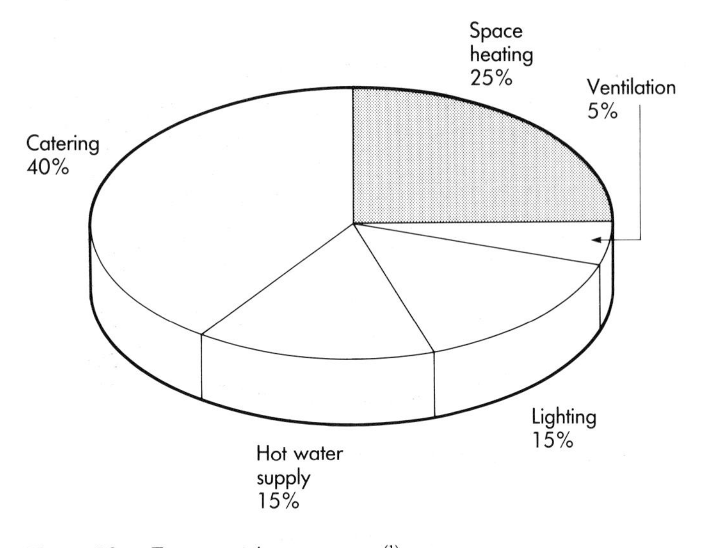

Figure 5.9 Energy use in a restaurant[1]

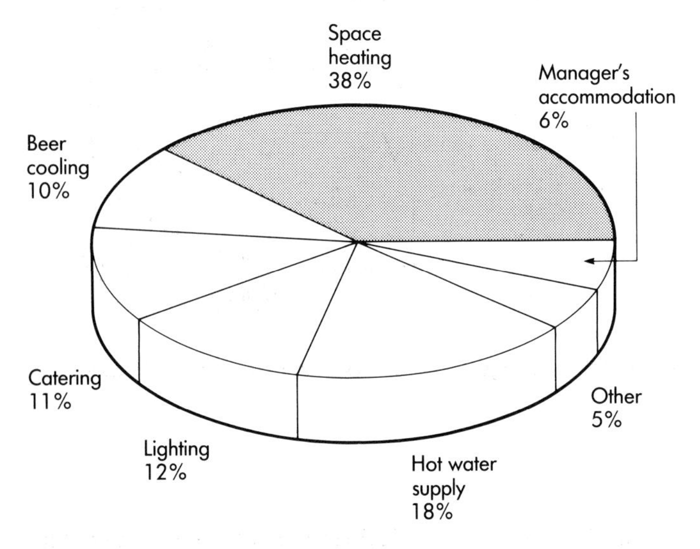

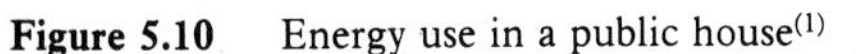
Figure 5.10 Energy use in a public house[1]

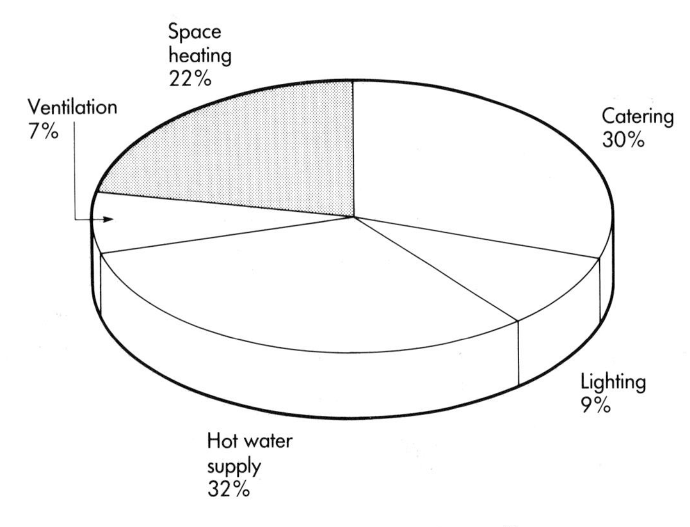

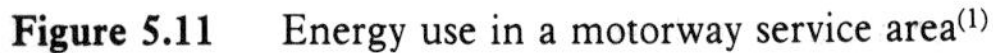
Figure 5.11 Energy use in a motorway service area[1]

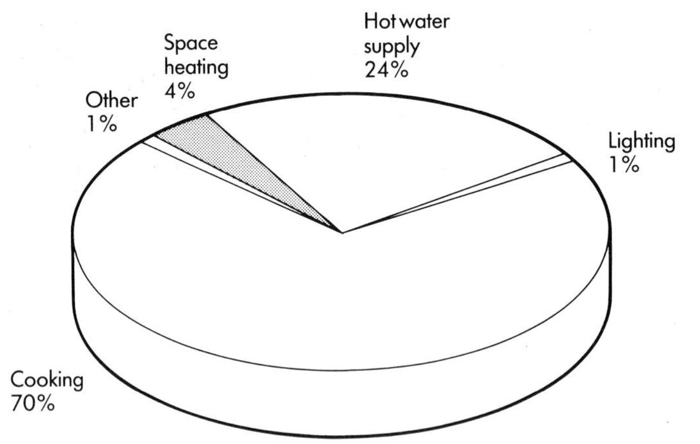

Figure 5.12 Energy use in a fast food outlet[1]

Shops

In contrast to the supermarket in Figure 5.13, smaller shops are likely to have no bakery and limited refrigeration. Larger stores may also have restaurant areas.

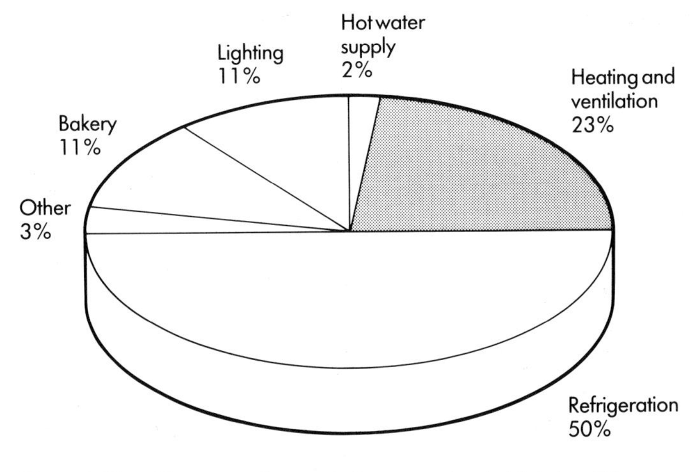

Figure 5.13 Energy use in a typical supermarket (with bakery)[1]

Offices

Figures 5.14 and 5.15 show how energy use can differ between naturally ventilated and air-conditioned offices.

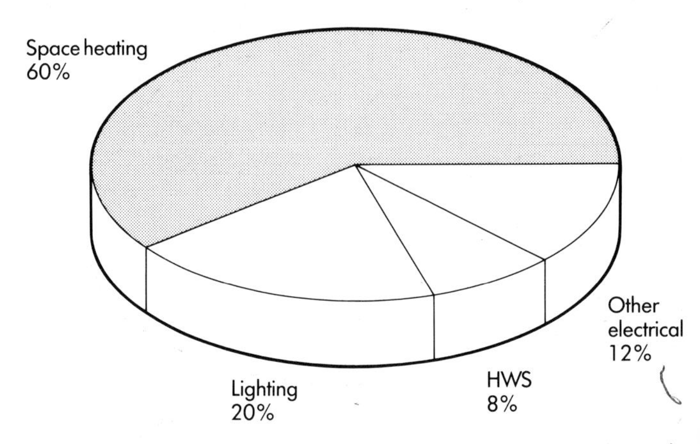

Figure 5.14 Energy use in a typical office building (naturally ventilated)[1]

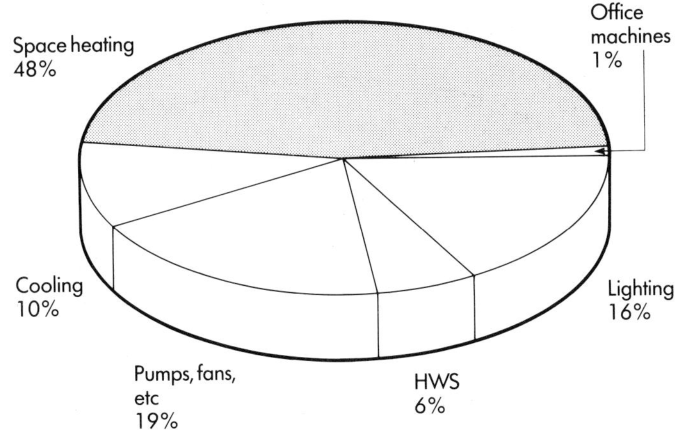

Figure 5.15 Energy use in a typical office building (air-conditioned)[1]

Sports centres and swimming pools

Figure 5.16 refers to a centre without a pool. A sports centre with a swimming pool that is more than 20% of the total floor area should be treated as a swimming pool, Figure 5.17. Pools equipped with full heat recovery should have a lower requirement for pool hall heating.

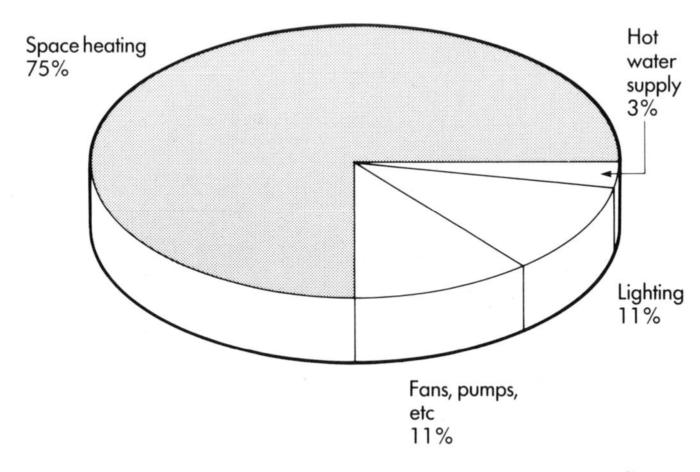

Figure 5.16 Energy use in a typical sports centre (without swimming pool)[1]

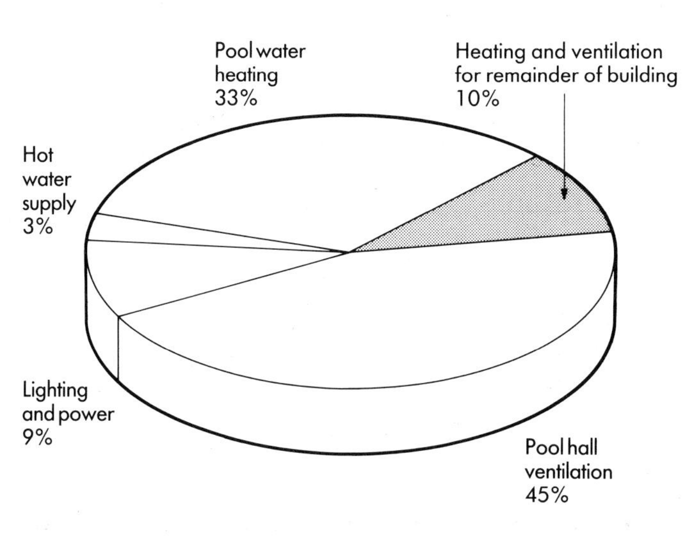

Figure 5.17 Energy use in a typical swimming pool[1]

Churches, libraries and galleries

In a church, Figure 5.18, energy is used predominantly for space heating. Other services, including air-conditioning and lighting, are significant in libraries and galleries, Figure 5.19, where specific conditions must be maintained.

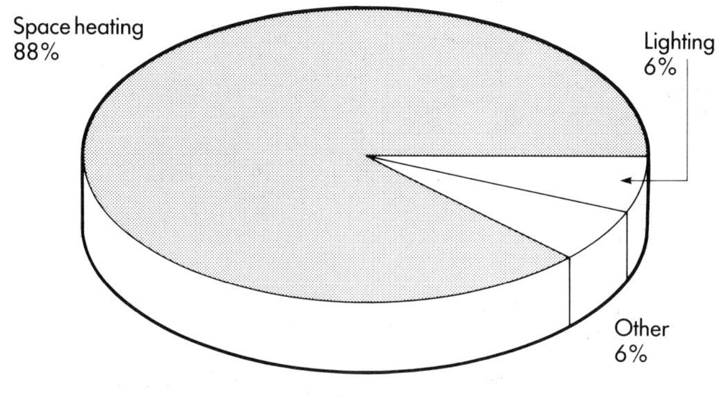

Figure 5.18 Energy use in a church[1]

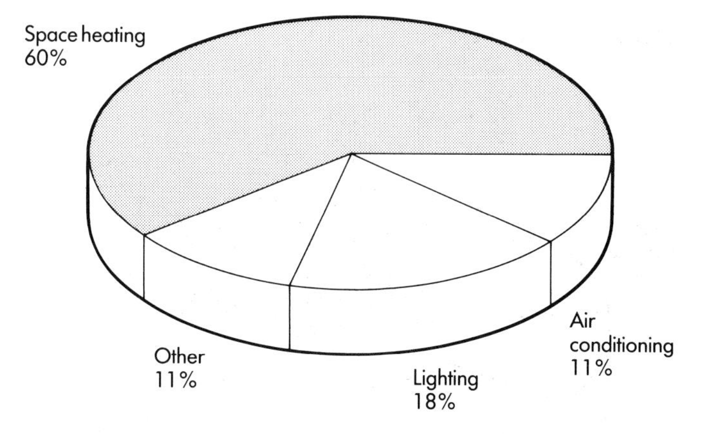

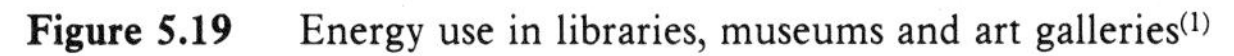

Figure 5.19 Energy use in libraries, museums and art galleries[1]

Hotels

Figure 5.20 refers to a typical large hotel. Extensive conference facilities may introduce air conditioning as a significant element in some hotels.

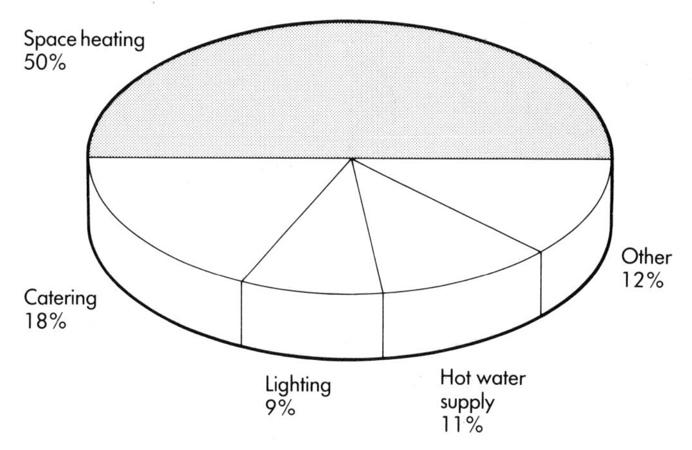

Figure 5.20 Energy use in large hotels[1]

Banks and agencies

Figure 5.21 refers to a bank without air-conditioning. If internal and solar gains demand the use of air-conditioning a reduced proportion for space heating should be expected. With increasing use of computers the provision of comfort cooling is now becoming the norm.

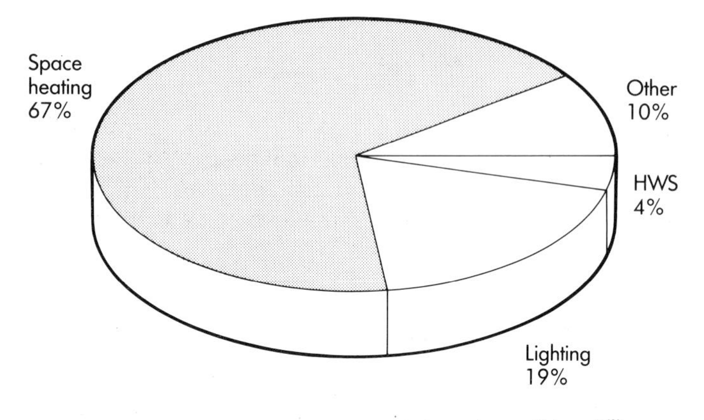

Figure 5.21 Energy use in a typical bank (non air-conditioned)[1]

Entertainment

For entertainment buildings such as cinemas and theatres, Figure 5.22, with high ceilings the proportions for space heating and ventilation are likely to be dominant. Cinemas also show a very low proportion for lighting. In clubs and halls, Figure 5.23, other uses must also be considered.

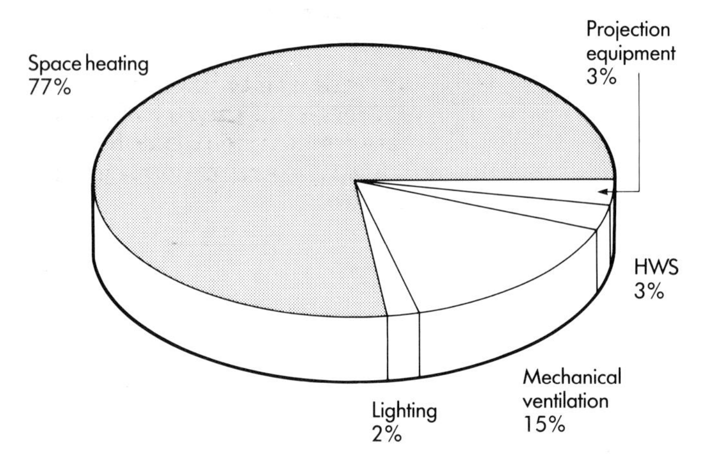

Figure 5.22 Energy use in a typical cinema[1]

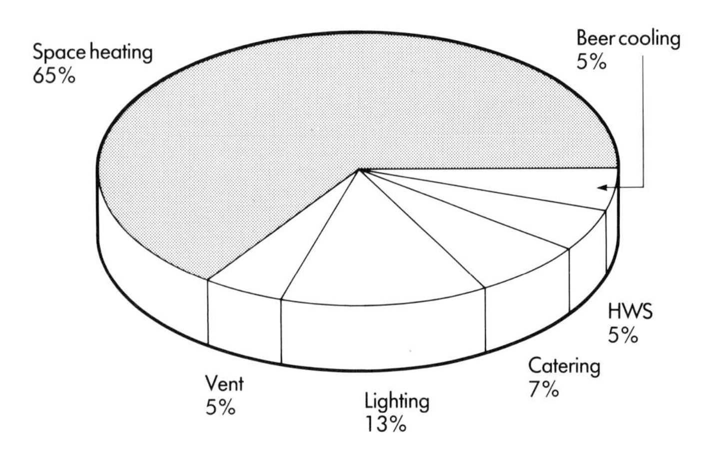

Figure 5.23 Energy use in bingo halls and social clubs[1]

Prisons

Figure 5.24 refers to the main services provided in a prison. The energy use for laundering or light industrial work might sometimes be identified separately if considered significant.

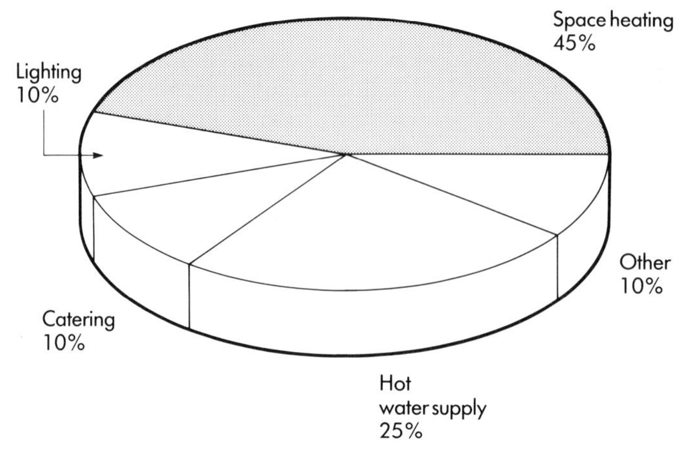

Figure 5.24 Energy use in a prison[1]

Transport depots and courts

Depots, Figure 5.25, and courts, Figure 5.26, both use energy mainly for space heating. Similar figures also apply to emergency services buildings.

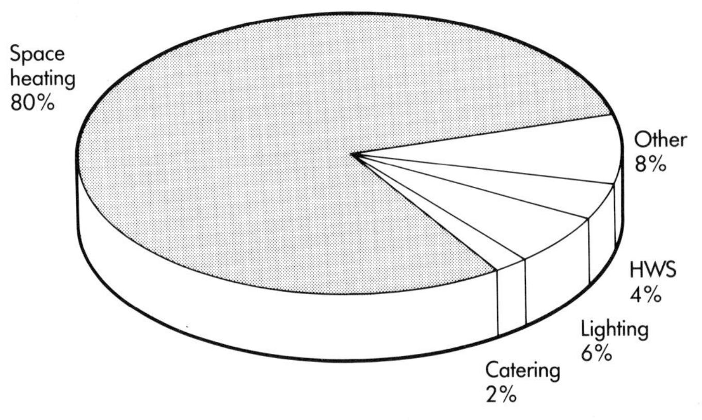

Figure 5.25 Energy use in transport depots[1]

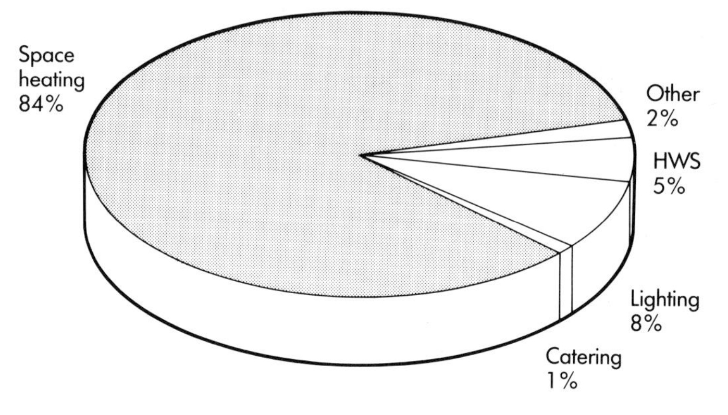

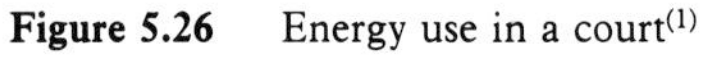

Figure 5.26 Energy use in a court[1]

Factories and warehouses

Process energy is excluded in Figure 5.27. Factories with large process heat gains should show a lower proportion for space heating but may have a significant ventilation requirement. Small, low-energy factory units may also have a lower proportion for space heating.

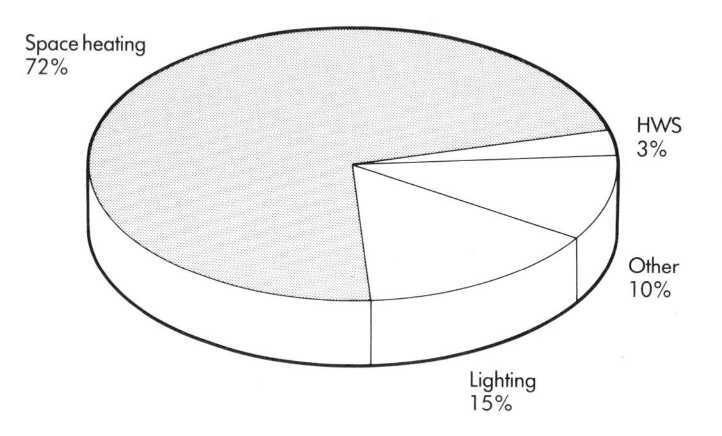

Figure 5.27 Energy use in a typical factory (excluding process energy)[1]

In a cold store, Figure 5.28, the main requirement is for refrigeration. If freezing down is also performed then the relevant proportion may be higher than that shown.

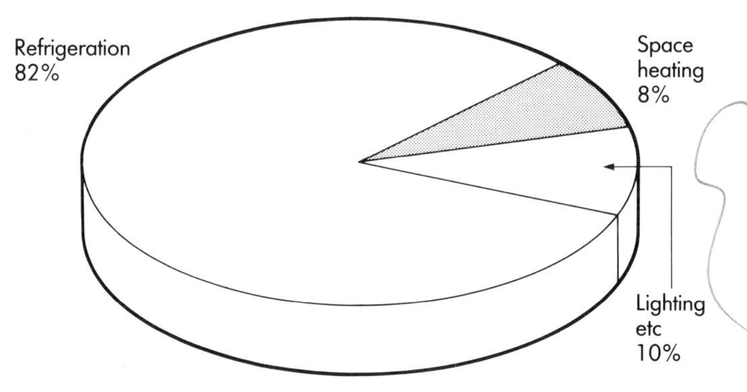

Figure 5.28 Energy use in a typical cold store[1]

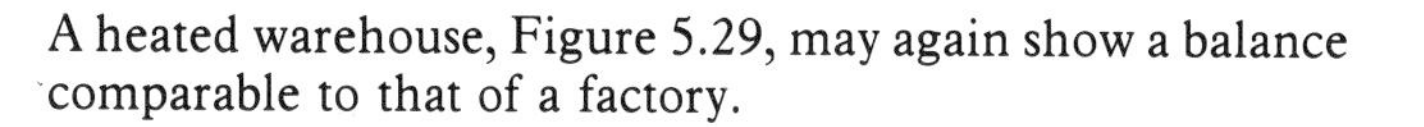

A heated warehouse, Figure 5.29, may again show a balance comparable to that of a factory.

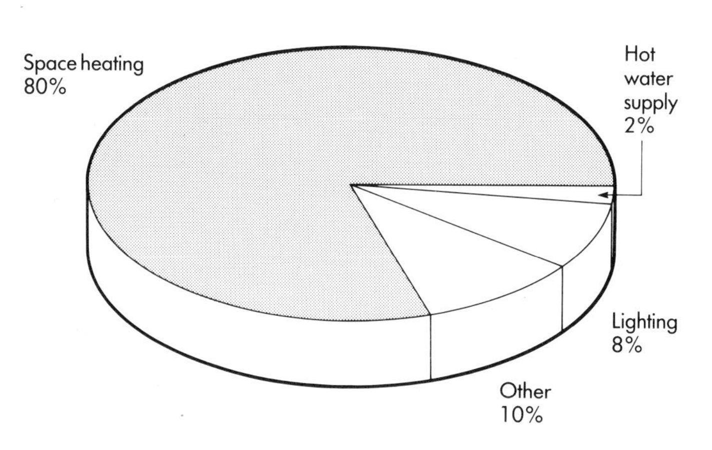

Figure 5.29 Energy use in a typical warehouse[1]

5.3 Deciding priorities for a site survey

Priorities should be determined from an assessment of the likely opportunities for savings, the value of energy involved and the available resources.

5.3.1 Single buildings

The services or end users identified as most significant in the breakdown of energy cost should normally be given priority in a survey. Priority can be given to areas where savings might be identified for the least cost and effort, e.g. simple space heating systems.

5.3.2 Multi-building sites

For buildings that are individually metered a comparison can be made both of energy costs and energy performance. If buildings are supplied by a common distribution system without sub-metering then energy yardsticks might be used to provide some guidance as to their likely energy use. Any major buildings that are obviously intensive energy users should be given priority. Other buildings may be selected on the basis of relative performance or on account of the types of service installed.

5.3.3 Multi-site organisations

Comparison of normalised performance indicators, combined with the actual size of energy bill, should indicate the sites to be given priority. Early consideration should be given to all sites with a very bad apparent energy performance and those with the greatest energy cost.

5.4 Summary of section 5

* Fuel invoices and site energy records covering at least one full year are required for analysis.
* Data should be checked carefully to ensure that they provide a complete record, have had any appropriate correction factors applied and can be identified with known supply points.
* Total energy intakes and costs should be determined and the relative proportions determined in both energy and cost terms.
* Performance indicators for total annual energy use can be calculated and compared with yardsticks for buildings of similar type to indicate whether good opportunities for savings are likely.
* Examination of monthly data can reveal seasonal patterns of use and base loads.
* Major energy users should be identified and evaluated where consumption records exist.
* Likely major users can be identified by reference to typical breakdowns of energy use for common building types.
* A cost breakdown can be derived from an energy breakdown and should be used to determine priorities for a site survey.

Reference for section 5

1 *Energy Efficiency in Buildings* (London: Energy Efficiency Office) (March 1989) 13 booklets

6 Planning a site survey

Where the preliminary audit dealt principally with the supply of energy, the site survey also examines each form of energy demand. The ways in which energy is converted, distributed and used are all potential subjects for study, as are the applied management procedures. A full investigation should deal with all of these elements to determine the best overall plan for improving efficiency and controlling energy costs. Studies concentrating on selected aspects of energy management or use should still take the possible wider implications into account.

To ensure that an effective survey brief can be developed consideration must be given to:

— who will be involved

— the building or site boundaries

— depth of survey required

— when the survey will be performed

— what further metering or instrumentation is necessary

— restrictions on access

— requirements for reporting.

6.1 Staff involvement

It is desirable that the project be directed by a person of managerial or board status to give authority to the survey and its findings. Whether outside assistance is used will depend principally on the complexity of the site and its buildings and systems and the availability of suitable staff. An energy manager, where in post, should be involved directly, but other interested parties representing end users and, perhaps, financial staff need to be consulted from the planning stage.

6.2 Site or building boundary

A single building such as an office block usually presents few problems as to the boundary of the survey. In cases of joint occupancy it is preferable to include the entire building if possible. Arbitrary divisions might otherwise be necessary, particularly where communal services are involved. Other sites must normally be considered as a group of buildings, cost centres, areas or zones, all of which have to be defined.

For sites with multiple buildings it is often preferable to specify each individual building to be included, particularly if buildings vary in construction and use. The alternative approach is to deal with particular services, such as space heating or lighting, and perhaps include each building or area supplied with the specified service.

It is important to identify any building or department which is to be excluded from the survey for some valid reason. For example, there may be a number of unheated stores, with minimal lighting which, although they occupy a significant area, have negligible energy consumption. Where there is initially some doubt, no area should be excluded until enough data have been collected to make an informed decision.

Useful building data should be collected at this stage:

— Prepare a schedule of sites or buildings and identify their main functions and total floor areas.

— For each building also identify the basic occupancy pattern and the main users of energy.

— Collect site or building plans showing basic dimensions and layouts. Drawings may be used to determine building sizes or in preparing building schedules or metering schematics. Major plant rooms can also be indicated to identify centres of energy conversion or distribution.

6.3 Depth of survey

The depth of study and the detail with which it is reported will be determined by the availability of resources and the value placed on the anticipated opportunities.

6.3.1 Comprehensive survey

A comprehensive survey is one that deals with specified areas or items in depth. It does not necessarily include every aspect of energy use but is annually expected to:

— determine the energy performance of buildings and major plant

— evaluate the principal energy flows

— identify precisely where savings can be made

— indicate the value of those savings

— provide costed recommendations or options as to how the savings can be achieved

— review procedures for energy management.

The report produced may, in addition to the recommendations and financial analyses, include:

— summaries of data collected

— details of rejected options

— outline or conceptual design work.

Appendix A4 summarises the expected scope of a comprehensive survey and is intended for use as a model brief.

6.3.2 Concise survey

A concise survey also deals with specific areas or items, but not in depth. It will:

— assess the energy performance of buildings and major plant using little or no direct measurement

— determine the principal energy flows

— identify the main opportunities for savings

— indicate the scale of those savings

— provide outline recommendations and costs.

The report should summarise the options and identify some opportunities for more detailed investigation. Appendix A5 summarises the expected scope of a concise survey and is intended for use as a model brief.

6.4 Survey timing

Careful timing of a survey will produce the best results. Programming should seek to take advantage of seasonal factors and other planned activities.

6.4.1 Seasonal factors

- Systems should generally be examined under operating conditions, i.e. heating in winter, cooling in summer, when performance can be measured. Some plant might be available for testing at any time.
- The preliminary audit may well indicate which of the heating and cooling loads is likely to offer the greater potential for savings. A summer and/or winter survey can then be selected as required.
- Examining plant after major maintenance is generally more appropriate than shortly before a shutdown. Ideally, examination both before and after will reveal the sensitivity of the equipment to maintenance.
- Observation of equipment not in operation can still provide much valuable information for a concise survey. For example, plant condition and control settings can be recorded. Details of model, type and rating could also be obtained from equipment. Remember that operating loads may well differ from design rating.

6.4.2 Programme

- Plan in conjunction with other proposed works, e.g. refurbishment or redevelopment.
- Choose start and completion dates that are convenient to the pattern of normal business.
- Select a period when key staff are available for consultation and not otherwise committed.
- Aim to disrupt normal business as little as possible, but do not confine the survey to periods when normal operating patterns cannot be observed.
- Plan around annual, weekend or overnight shutdown periods. These might be periods when no useful site work can be performed but might, conversely, allow access or observations not possible during normal operation. Base loads and standing losses might best be observed during 'quiet' periods.
- Assess the necessary duration of site investigation, particularly where measurements are required over some fixed period, e.g. load patterns or temperature profiles over one week.
- Determine not only when site work is to be performed but also when the report and recommendations have to be available, for instance to meet financial deadlines.

The precise timing of the survey will be affected by the availability of resources — staff, funding and data. If existing records of consumption are limited, perhaps because site meter readings are not maintained, it may be preferable to defer the survey until more comprehensive records can be established. Additional data may be invaluable when assessing the breakdown of energy use.

Once the programme has been finalised, time limits should be set and adhered to in order to maintain momentum.

6.5 Additional metering

It may be evident from the outset that additional meters installed at strategic points could provide much additional information. A period may be required before the survey to collect a useful number of readings from new meters or existing meters that have not previously been read.

Figure 5.3 illustrated how schematic diagrams of meter points in any energy system might well highlight anomalies such as an imbalance between the consumption recorded by a main meter and the total of sub-metered consumptions. Shortcomings in the metering arrangements can also become evident once the end users without sub-metering have been identified or where several are served by a single meter. The use of portable or temporary equipment for cross-checking old or disused meters should be considered.

Permanent metering should be considered where the cost of hiring, installing and removing temporary meters may be more than the purchase cost. This may apply particularly to electrical meters. The inconvenience of shutting down energy supplies for meter installation must also be considered; shutdowns might only be possible during certain limited occasions if essential services cannot be interrupted.

6.6 Access to site

Restraints may be imposed on survey staff and working practices. Any of the following items should be identified:

- hazardous areas, e.g. high voltage, dangerous substances
- requirements for approved or authorised staff, e.g. qualified to work on high-voltage equipment
- security clearance requirements
- medical health requirements
- need for protective clothing
- restrictions on use of survey instruments, e.g. to intrinsically safe instruments operating on low voltage and containing no toxic materials
- safety procedures.

Department heads and security staff should be notified of the programme and asked to co-operate in the smooth running of the survey.

6.7 Reporting requirements

The immediate end product of a survey is a report. Reporting procedures must therefore be considered at an early stage. Note that the effort involved in evaluating findings and preparing a final report is normally at least as great as that spent on site work.

6.7.1 Content

Particular requirements and expectations should be established before site work commences. A management summary is recommended for presentation to senior management.

6.7.2 Interim reports

Periodic verbal and written progress reports can allow early implementation of simple measures during extended periods of site work. They can also help to ensure that effort continues to be applied most effectively.

6.7.3 Draft report

Circulation of a draft version of the final report is recommended, particularly in the case of a major report prepared by an outside organisation. Additional time must be allowed for discussion and amendments.

6.7.4 Final report

The date specified for completing the final report may determine the whole programme. Formal presentation is recommended.

6.7.5 Formal presentation

The final report should be presented to the level of management with authority to implement the capital schemes recommended.

6.8 Summary of section 6

- When planning, consider who will be involved, what is to be examined, when the survey will be performed and what further resources might be necessary.
- The authority of senior management and the commitment of all staff should be obtained at the outset.
- Site boundaries must be defined by specifying individual buildings or areas to be included and those which are to be excluded.
- The depth of survey must be decided — this could be a brief examination or a detailed investigation.
- Careful timing in respect of operating patterns and shutdowns can help to make the survey more effective.

7 Site survey — energy management and operational requirements

A survey should normally include an examination of the effect that operations have on energy demand and the management of energy in meeting that demand.

7.1 Energy management

Management practices applied in the monitoring and control of energy purchase and consumption must be reviewed. The involvement of staff at all levels in an organisation should be reviewed with reference to:

- **accountability** from end use to overall financial control
- **procedures** adopted in respect of management information
- **maintenance** insofar as it affects energy use.

7.1.1 Accountability

As with any aspect of business there should be clear accountability or responsibility for energy. The organisation of the relevant management structure needs particular consideration:

- Assess the number of full-time equivalent staff with specific responsibility for energy management against the total cost of energy purchased. A guideline of one full-time equivalent energy manager for each £1 million expenditure on energy is recommended for local authorities by the Audit Commission[1]. Even if the energy bill justifies only limited time expenditure by someone with little specialist knowledge there should still be formal recognition of the duty. Relevant skills and experience are ex-pected if the post is full-time.
- Consider the position of those responsible for energy management in an organisation. If the role is merely advisory then there must be a mechanism for reporting to someone with the authority to act.
- Identify areas, or cost centres, where those accountable for operating costs can be made individually accountable for energy use. Individuals with control of budgets should be given targets for energy use and held responsible for overspend on energy. Responsibility should extend to ensuring that the occupants and equipment under their control use energy efficiently.
- Examine how the task of reading meters and collecting other routine monitoring data is performed. This must be another recognised duty to ensure that information is supplied reliably.
- Review the level at which energy-related matters are discussed. Those with ultimate financial control of budgets must have a declared policy that permits investment in energy conservation.
- Assess the training, motivation and incentives needed at all levels to encourage staff involvement.

7.1.2 Procedures

Good management relies on the effective collection, processing, dissemination and use of information. Regularly implemented procedures must be examined in detail to ensure that the correct information is made available at the right time and in the right place. This requires the operation of a monitoring and target setting system.

- Assess the adequacy of the metering arrangements and the methods and frequency of recording consumption and the principal variables which affect it. Recommend additional metering or readings where necessary.
- Review the methods of analysing data which could include:
  - (*a*) application of correction/conversion factors and use of standard units
  - (*b*) allocation to cost centres/end users
  - (*c*) use of standards and targets for performance comparison.
- Examine the format used to record results. It should indicate clearly when and where action is needed.
- Review the use made of the results or findings. Timely information must get to those in a position to take action. Check that routes of communication are clearly established and used.
- Confirm that the management structure exists to co-ordinate effort and ensure that action is taken when necessary.

7.1.3 Maintenance

The maintenance of plant, equipment and buildings has a significant effect on energy use. It is an important issue that will normally justify a separate study. Detailed guidance can be obtained from specialist sources.[2]

For the purposes of an energy survey it is necessary to:

- recognise where the frequency or quality of maintenance is affecting energy use adversely
- identify those energy efficiency measures which should be implemented as maintenance items under separate budgets.

Reliable operation of some essential plant is not always seen as being necessary to optimum energy efficiency. A good management policy recognises that efficient maintenance can help to ensure efficient operation. Adequate maintenance resources are vital to effective energy management. An energy survey should highlight problems caused by deficiencies in this area but is not expected to report on maintenance management in depth.

7.2 Operational requirements

Operational requirements reflect management policy and determine the true demand for energy. Reduction in this demand should lead to a direct saving in purchased energy. Changes in policy may be suggested where needed to allow such a reduction. Improved knowledge of the requirements can also assist efficient matching of the supply and demand, with services supplied only as necessary.

Details of the operational parameters that should be established include:

— building use
— occupancy patterns
— number and type of occupants
— operating patterns of major processes
— specified environmental conditions for occupants, equipment and buildings
— basic services.

Standard sheets, of which Table 7.1 is an example, can be used for each building, area or zone, as appropriate, for ease of reference during the survey.

Table 7.1 Building data sheet

Building	Name/Number	:
Function(s)	– Category	:
	– % of area	:
Total floor area	– (m^2)†	:
No. of floors		:
Volume	– (m^3)†	:
Occupancy	– No/Type of occupants	:
	– Days/week	:
	– Daily times	:
Environmental requirements (°C, %RH, etc.)	– During occupancy	:
	– Other periods	:
Building fabric	– Date built	:
	– Wall construction	:
	– % glazing	:
	– Roof	:
	– Floor	:
Services supplied (state uses)	– Electricity	:
	– Natural Gas	:
	– Oil	:
	– Coal	:
	– Steam	:
	– HTHW/LTHW	:
	– Compressed air	:
	– Other	:
Major plant items		:

†State percentage heated.

7.2.1 Building use

The suitability of a building for its function and the way in which it is used affect its energy requirements.

— Establish all the main uses of a building and the approximate proportions of the total area that they occupy.

— Note changes from original design function; systems may have become inappropriate with a change of use, e.g. an area of an air-conditioned office converted for storage no longer requires fully treated air.

— Examine combinations of use with complementary or opposing heating/cooling needs. Heat transfer between areas may be appropriate. Alternatively, further physical separation might be required, e.g. locate heat producing equipment away from an air-conditioned office.

— Consider grouping functions with similar requirements and hours of use in specific areas, buildings or sites. This could allow rationalisation of site distribution services, e.g. relocation of all steam users to a few buildings on an extensive site could allow isolation of steam supply to all other buildings, with a consequent saving in distribution losses. There may be a case for concentrating heat producing equipment in rooms on the shaded north side of a building to restrict the need for comfort cooling.

7.2.2 Occupancy patterns

— Establish the patterns of occupancy in each area or building and relate them to the installed systems and controls. Note where zoning of services could be improved to match the occupancy or where the location and working patterns of the occupants might be altered instead. With flexitime working, heating could perhaps be restricted to a basic core time.

— Identify where systems lack flexibility of control and note where it may, for example, be possible to restrict out-of-hours use to particular areas.

— Review the extent to which occupants are able to control their own local environment. Accessible controls can encourage good housekeeping but result in undesirable tampering.

7.2.3 Occupancy

The number and type of occupants are particularly relevant to ventilation requirements, fresh air supply, air change rates, temperature and humidity.

— Establish the type of occupants; human and/or animal; active/inactive; smokers/non-smokers.

— Check the number of occupants; fixed or variable; same occupants or changing.

— Determine the environmental requirements.

Some typical basic design conditions are given in Table 7.2[3]. For most purposes, relative humidity between 40% and 70% will be satisfactory for comfort.

Table 7.2 Design conditions for UK[3]

Season	Occupancy category	Resultant temperature (°C)	Relative humidity (%)
Summer	Continuous	20–22	50
	Transient	23	50
Winter	Continuous	10–20	50
	Transient	16–18	50

Recommended outdoor air supply rates for various types of air conditioned spaces are given in Table 7.3[3], and Table 7.4[4] gives recommended total air supply rates.

Table 7.3 Recommended outdoor air supply rates for air-conditioned spaces [3]

Type of space	Smoking	Outdoor air supply (litre/s)		
		Recommended	Minimum (Take greater of two)	
		Per person	Per person	Per m² floor area
Factories †‡	None			0.8
Offices (open plan)	Some			1.3
Shops, department stores		8	5	
and supermarkets	Some			3.0
Theatres †	Some			—
Dance halls †	Some			—
Hotel bedrooms ‡	Heavy			1.7
Laboratories ‡	Some	12	8	—
Offices (private)	Heavy			1.3
Residences (average)	Heavy			—
Restaurants (cafeteria) ‡¶	Some			—
Cocktail bars	Heavy			—
Conference rooms (average)	Some	18	12	—
Residences (luxury)	Heavy			—
Restaurants (dining rooms) ‡	Heavy			—
Board rooms, executive offices and conference rooms	Very heavy	25	18	6.0
Corridors	A *per capita* basis is not appropriate to these spaces.			1.3
Kitchens (domestic) ‡				10.0
Kitchens (restaurant) ‡				20.0
Toilets †				10.0

† See statutory requirements and local bye-laws.

‡ Rate of extract may be over-riding factor.

¶ Where queuing occurs in the space, the seating capacity may not be the appropriate total occupancy.

Notes

1 For hospital wards, operating theatres see *Department of Health Building Notes.*

2 The outdoor air supply rates given take account of the likely density of occupation and the type and amount of smoking.

Table 7.4 Recommended total air supply rates for various building types[4]

Type of building or space	Recommended total air supply rate (air changes/hour)
Banking halls	6
Canteens	8–12
Cinemas	6–10
Dining halls, restaurants	10–15
Kitchens, hotel and industrial	> 20
Laboratories	4–15
Laundries	10–15
Lavatories and toilets	> 5
Libraries, museums, galleries	3–4
Offices	4–6

For space heating, the maximum temperature in premises as set by the 1980 *Fuel and Electricity (Heating) (Control) (Amendment) Order*[5] is 19°C. The *Factories Act*[6] of 1961 and the *Offices, Shops and Railway Premises Act 1963*[7] specify minimum temperatures of 15.5°C and 16°C respectively to be reached within one hour of occupation.

Recommended design values[3] for dry resultant temperature are given in Table 7.5. The resultant temperature is the temperature recorded at the centre of a blackened globe 100 mm in diameter. Air temperature is typically 1–2°C below resultant temperature with radiant heating systems and 1–2°C above resultant temperature with convective heating systems. Temperatures within ± ½°C of the chosen value should cause no significant increase in dissatisfaction.

Table 7.5 Recommended design values for dry resultant temperature[3]

Type of building	t_{res} (°C)
Art galleries and museums	20
Assembly halls, lecture halls	18
Banking halls	
Large (height > 4 m)	20
Small (height < 4 m)	20
Bars	18
Canteens and dining rooms	20
Churches and chapels	
< 7000 m^3	18
> 7000 m^3	18
Vestries	20
Dining and banqueting halls	21
Exhibition halls	
Large (height > 4 m)	18
Small (height > 4 m)	18
Factories	
Sedentary work	19
Light work	16
Heavy work	13
Fire stations, ambulance stations	
Appliance rooms	15
Watch rooms	20
Recreation rooms	18
Flats, residences, and hostels	
Living rooms	21
Bedrooms	18
Bed-sitting rooms	21
Bathrooms	22
Lavatories and cloakrooms	18
Service rooms	16
Staircases and corridors	16
Entries halls and foyers	16
Public rooms	21
Gymnasia	16
Hospitals	
Corridors	16
Offices	20
Operating theatre suite	18–21
Stores	15
Wards and patient areas	18
Waiting rooms	18

Type of building	t_{res} (°C)
Hotels	
Bedrooms (standard)	22
Bedrooms (luxury)	24
Public rooms	21
Staircases and corridors	18
Entrance halls and foyers	18
Laboratories	20
Law courts	20
Libraries	
Reading rooms (height > 4 m)	20
(height < 4 m)	20
Stack rooms	18
Store rooms	15
Offices	
General	20
Private	20
Stores	15
Police stations	
Cells	18
Restaurants and tea shops	18
Schools and colleges	
Classrooms	18
Lecture rooms	18
Studios	18
Shops and showrooms	
Small	18
Large	18
Department store	18
Fitting rooms	21
Store rooms	15
Sports pavilions	
Dressing rooms	21
Swimming baths	
Changing rooms	22
Bath hall	26
Warehouses	
Working and packing spaces	16
Storage space	13

7.3 Summary of section 7

- Energy management practices must be reviewed with particular reference to accountability and procedures.
- Maintenance has a significant effect on energy efficiency.
- There should be a management structure with clear responsibility for energy and the authority to act.
- Individuals should be identified and made accountable for energy use in specific cost centres.
- The need for training and motivation at all levels must be recognised.
- Good management requires regular and reliable energy data.
- Metering arrangements and data analysis techniques should be reviewed.
- Mechanisms must be identified to ensure that the right information gets to the right people at the right time to act.
- The operational requirements determined by business needs or management policy directly affect energy demand.
- Changes in the way that a building is used can lead to a reduction in energy demand.

References for section 7

1 *Audit Guide on the Management of Energy in Local Authority Buildings* (London: The Audit Commission in collaboration with the Energy Efficiency Office)

2 *Maintenance management for building services* Technical Memoranda **TM17** (London: Chartered Institution of Building Services Engineers) (1990)

3 Environmental criteria for design *CIBSE Guide* Section A1 (London: Chartered Institution of Building Services Engineers) (1986)

4 Ventilation and air conditioning (requirements) *CIBSE Guide* Section B2 (London: Chartered Institution of Building Services Engineers) (1986)

5 *Fuel and Electricity (Heating) (Control) (Amendment) Order 1980* SI 1013 (London: HMSO) (1984)

6 *Factories Act 1961* (London: HMSO) (1961)

7 *Offices, Shops and Railway Premises Act 1963* (London: HMSO)(1963)

8 Site survey — energy supply

The arrangements for the supply of each form of energy must be examined not only individually but also collectively to determine the best overall supply policy. Changes in the structure of the supply industries following privatisation increase the need for regular review of the purchase of energy.

8.1 Electricity

The principal sources of supply are:

- Generating/distribution companies
- Landlord supply
- On-site generation (in parallel with grid supply or as standby).

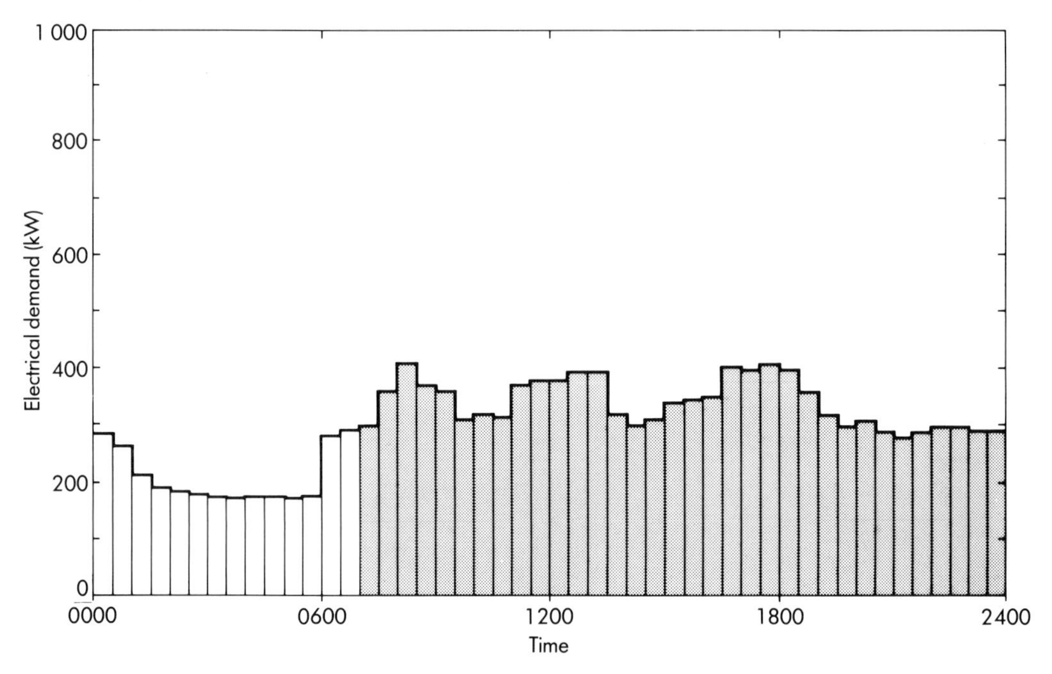

Figure 8.1 Example of electrical load pattern. Tariff 1 operates during the period midnight to 0700. Tariff 2 operates during the period 0700 to midnight.

8.1.1 Generating/distribution companies

Companies supplying electricity base their charges on the use measured at the point of supply. It is here that any measures to be implemented must have an effect. If charges are based only on unit consumption, and there are no tariff options, then priority should be given to controlling consumption. When demand charges are also imposed, any study should also examine the load pattern and power factor. Large users may be able to negotiate a supply contract with a generating company rather than with their local distribution company.

8.1.2 Landlord supply

A landlord supply should normally be metered and charged at a set rate. Alternatively, a fixed service charge may be applied, perhaps in proportion to the area served. There may still be scope for renegotiating the unit cost or service charge if examination of the load pattern and consumption suggest that this might be appropriate. For example, a high proportion of off-peak use could be grounds for requesting a reduced charge.

8.1.3 On-site generation

Generating costs need special consideration in respect of the accounting procedures applied to each aspect of operation and maintenance. This may have to be made the subject of a separate study.

8.1.4 Load pattern

Review the load pattern at each point of supply over a period of at least one day and preferably one week:

- For minor supplies take readings at the beginning and end of the selected period and at intermediate times to indicate overnight or weekend use as appropriate. Many useful conclusions can often be drawn from a few such readings, e.g. the consumption during the night could suggest that a change of tariff is worthwhile.
- For major supplies and particularly those on maximum demand tariffs, use a temporary portable recorder to record consumption in kWh or kVAh at 30-minute intervals instead of taking intermediate readings.

Figure 8.1 illustrates how a load pattern obtained by recording the load in kW at half-hourly intervals can show the consumption in different tariff periods and the size and timing of peak loads. Morning, midday and early evening peaks can be seen in this example.

- Note when peak loads occur and try to identify what causes them. Further site investigation may show how peaks can be prevented or reduced to minimise maximum demand charges where applicable.

For charging purposes, the maximum demand recorded in any month is normally based on the maximum unit consumption in any 30-minute period during that month. To keep this to a minimum the survey must therefore seek opportunities to minimise maximum demand by load shedding, i.e. switching off major items of plant for periods to avoid peaks. Large fan motors are often selected for manual or automatic load control where ventilation can be interrupted without serious inconvenience. If suitable plant items cannot be identified then an equivalent result might be achieved by a general reduction in the base load. Methods of base load reduction include good housekeeping, improvement of power factor and more efficient lighting systems. Table 8.1 is a worked example of the annual saving that might be achieved with a reduced base load, based on the appropriate published demand charges. It may be useful to assess the principal components of the base load, as in Figure 8.2, to indicate possible opportunities for reduction.

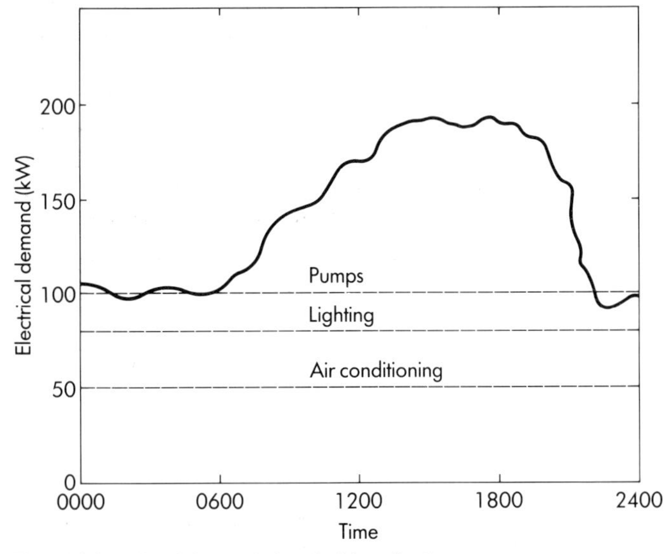

Figure 8.2 Breakdown of electrical base load

Table 8.1 Example calculation of the annual saving available from a general reduction of 20 kVA in the maximum demand in a building with a monthly maximum demand tariff.

Month	Published demand charge (£/kVA/month)	Calculated saving assuming 20 kVA reduction in demand (£/month)
April–Oct	Nil	Nil
Nov	2.24	44.80
Dec	7.14	142.80
Jan	7.14	142.80
Feb	2.24	44.80
Mar	Nil	Nil
	Annual saving	375.20

8.1.5 Tariff

Confirm the tariff applicable to each supply identified during the audit and consider the alternative tariffs that are available. Note that the site survey findings may suggest changes in the future load pattern and consumption which will have to be taken into account. The following controllable factors may have to be considered in deciding the most favourable tariff:

- total unit consumption
- day/night/evening/weekend/seasonal use
- agreed supply capacity or availability
- maximum demand (kW/kVA)
- power factor
- board metering at high voltage (normally 11 kV) or low voltage (normally 240 V or 415 V).

Sites with multiple supplies may offer scope for negotiating the combination of several supplies into a single account to take advantage of tariff structure.

A simple example of the saving available by a straightforward change of tariff to take advantage of a building's consumption pattern is given in Table 8.2. The minimum proportion of units that must be purchased at the night rate to make a change to the day/night tariff favourable in Table 8.2 can be calculated as:

$$\frac{(4.70 - 4.48)}{(4.70 - 2.08)} \times 100 = 8.4\%$$

8.1.6 Power factor

Determine the power factor of supplies to which demand charges apply either from readings of kWh/kVAh/kVArh meters if available or by direct measurement. Power factor over a period is calculated as PF = kWh/kVAh or cos ϕ, where tan ϕ = kVArh/kWh. The instantaneous value can be given as kW/kVA.

This relationship is represented in Figure 8.3. The necessary information may sometimes be available from electricity invoices, e.g.

Units recorded for month = 17 400 kWh
8 700 kVArh

Table 8.2 Example calculation of the annual saving available from a change of tariff for a building supplied under a standard monthly maximum demand tariff

Data	: Annual consumption	= 460 000 kWh
	: Proportion consumed during night period (2400–0700 h GMT)	= 18%
	: Tariff A (24 h rate) unit cost	= 4.48p/kWh
	: Tariff B (day/night rates) unit	= 4.70/2.08p/kWh
Annual unit charges under tariff A		= 460 000 × 0.0448
		= £20 608.00
Annual unit charges under tariff B		= 460 000 × 0.0470 × (1–0.18) + 460 000 × 0.0208 × 0.18
		= £19 450.64
Annual saving by changing from tariff A to B (assuming standing and demand charges are unchanged)		= £1 157.36

In this case

$$\tan \phi = 8700/17\,400 = 0.5$$

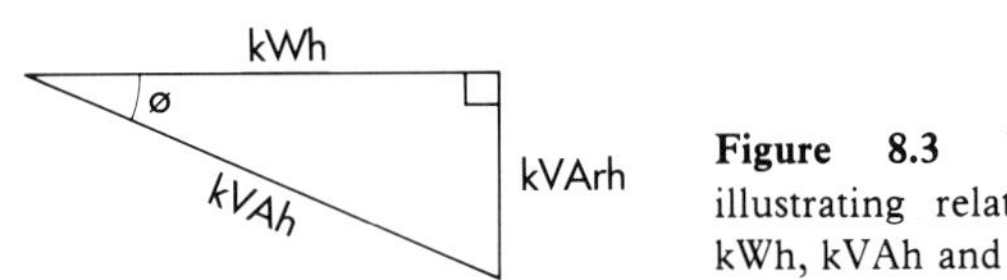

Figure 8.3 Vector diagram illustrating relationship between kWh, kVAh and kVArh

therefore power factor $\cos \phi = 0.89$. A power factor below this value, when kVArh units exceed 50% of kWh units, will be penalised by many supply companies. Principal causes of low power factors are induction motors and fluorescent lights.

Power factor correction must be given particular consideration where demand charges are based on maximum recorded kVA, rather than kW. The cost of installing corrective capacitance, either in bulk at the supply point or locally on major equipment, is often justified by the consequent reduction in charges. Depending on supply company, the recommended value of power factor to be maintained could vary between 0.85 and 0.98 (lagging). Each case must therefore be assessed on its merits. It should be noted that the return on investment is best when improving from a low value to a moderate value, say 0.80 to 0.93, and diminishes as unity is approached.

Table 8.3 shows the amount of capacitance needed for power factor correction in the range commonly experienced. The actual kW load multiplied by the figure from the table gives the kVA of capacitance needed.

Table 8.3 Size of capacitors in kVAr per kW of load needed to raise power factor

Existing PF	Required power factor				
	0.80	0.85	0.90	0.95	1.0
0.95	—	—	—	—	0.320
0.90	—	—	—	0.155	0.328
0.85	—	—	0.136	0.291	0.484
0.80	—	0.130	0.266	0.421	0.750
0.75	0.132	0.265	0.398	0.553	0.882
0.70	0.270	0.400	0.536	0.691	1.020

The satisfactory operation of existing power factor equipment should be checked — it may for example have been left switched out of service or be inoperative for some other reason.

8.2 Natural gas

Most consumers are now supplied under a standard published tariff. Those using less than 25 000 therms per annum have little or no choice of tariff. Larger users are charged at a scheduled reference price. The basic 'schedule price' is determined by:

— volume of consumption (in bands)

— number of premises included in contract (in bands)

— contract type.

The lowest price applies to single premises in the highest consumption band. For a surcharge the consumer has the option of a fixed price for the contract period i.e. one or two years. Seasonal pricing factors applicable to firm gas supplies reduce prices outside the winter period. Cheaper interruptible supplies are available to major consumers. Large users may also be able to obtain a contract with an independent gas supplier.

Contract details should be studied to ensure that gas is purchased under the most favourable terms:

(*a*) For premises with multiple supplies, determine whether there may be any advantage in combining supplies for metering purposes.

(*b*) For organisations with multiple premises, each consuming more than 25 000 therms per annum, check how premises are grouped or consumptions are aggregated for contract purposes.

(*c*) For premises consuming more than 250 000 therms per annum under a firm supply contract, examine the opportunities for interruptible operation; additional facilities for use of a standby fuel may be needed, depending on the possible period of interruption.

If there is no sub-metering then the load of a major user can sometimes be assessed by taking readings from the mains supply meter during periods when no other use is occurring.

8.3 Oil

Oil prices are subject to frequent fluctuation so costs are likely to differ from one delivery to the next. Large organisations should have the advantage of negotiating favourable contracts whereas smaller individual consumers may have flexibility in their choice of supplier.

Storage and handling facilities must be examined as they not only affect purchasing arrangements but can also be direct energy consumers.

(*a*) Consider the storage volume in relation to average and peak consumption rates. With residual grades of oil it may be possible to reduce the volume kept heated to reduce standing losses. It is normal to provide enough storage for approximately three weeks operation but this depends on individual circumstances. Additional capacity may enable those with interruptible gas supplies to operate on oil for longer periods and negotiate a cheaper gas price.

(*b*) Tanks in which heavy fuel oil is kept heated should be checked for adequate insulation and correct control of storage and temperatures. Weatherproofed insulation with a minimum thickness of 50 mm of glass wool or foam and a minimum storage temperature of 40°C is recommended for Class G fuel oil. Oil in tanks not needed for immediate service can be kept at a lower temperature, subject to standby requirements and the advice of the supplier. Seasonal reduction of the storage capacity kept heated may be possible.

(*c*) For residual oils, the oil temperature at the pump should be checked. A pumping temperature of 55°C is recommended for Class G fuel oil, maintained by an outflow heater.

(*d*) Determine whether a change of oil grade is worthwhile — the lower price of heavier grades must be set against the additional handling costs and account taken of all factors affecting operation. When comparing grades[1] of oil it should be noted that the heavier the grade, the higher is its calorific value per unit volume. The comparison should therefore be based on cost per gigajoule rather than cost per litre.

8.4 Coal

The cost of coal is mainly dependent on grade, with adjustments for location and quantity delivered. Bulk purchase contracts require the provision of a stockpile, bunker or silo to accept deliveries. Note that in the first few months following delivery there is usually a slight reduction in calorific value due to oxidation. After approximately six months any further reduction should be insignificant.

Investigations should concentrate on:

- the possibility of using alternative grades of coal with lower overall purchase, handling and operating costs. The system of transfer to the point of use and final de-ashing have a major effect on handling costs.
- effect of storage facilities/capacity on size and cost of deliveries
- procedures for using fresh/old stocks to minimise effect of reduction in calorific value.

Methods of recording stock levels and actual use should be examined for accuracy when considering monitoring procedures.

When coal consumption and deliveries are monitored on a mass basis it should be noted that significant errors can occur due to changes in moisture content. For example a coal may be delivered with 10% moisture, stored under cover (or in a silo) and then weighed (after drainage) with 5% moisture. Conversely, coal which is stored in the open could gain moisture before weighing. Ideally samples of coal should be checked for moisture content and monitored say on a 5% moisture or dry basis. When sampling[2] from a stock pile, the number and location of samples taken should aim to provide a representative result for the coal being used.

8.5 Liquefied petroleum gases (LPG)

LPG, usually propane or butane, is supplied either in bottles or in bulk to a storage tank. In both cases, the weight of delivery is given. However, gas consumption may be monitored using in-line gas volume meters with conversion to mass via density, including temperature and pressure correction factors.The use of direct-fired appliances within buildings is sometimes proposed as a convenient solution to some heat supply problems, provided that safety and potential condensation problems are not overlooked. Safe storage of LPG must be ensured at all times.

For all but very minor users the tank rental charges for bulk purchase are considerably less than the additional cost of bottled gas.

8.6 Other fuels

In addition to electricity and the main primary fuels there is a range of other sources of energy which may already be in use or could be considered, including:

- incineration of solid waste
- burning of waste oil
- passive/active solar energy
- waste process heat.

Further sources, such as wind energy, might also be available in some locations.

The survey should establish the practicality and costs of these methods or their potential use where a new opportunity clearly exists. Factors to be considered include:

- availability of supply
- type or grade of energy produced
- coincidence with demand
- cost of equipment
- value or disposal cost of fuel source compared with cost of fuel being replaced
- handling/processing costs
- storage requirements
- environmental impact
- need for plant attendants
- maintenance and other operating costs.

8.7 Summary of section 8

* Changes in the structure of the energy supply industries are increasing the need for regular review of the policy for purchasing energy.
* For all fuels lower prices may be possible by changing supply, delivery or storage arrangements.
* Electricity supply arrangements determine the types of savings measure that need to be identified, e.g. tariff changes, load control.
* Large gas consumers, particularly those with multiple sites, should identify the most favourable type of supply contract available to them.
* The feasibility of alternative, cheaper fuels should be investigated.

References for section 8

1 *BS 2869: 1983: Specification for fuel oils for engines and burners for non-marine use* (London: British Standards Institution) (1983)

2 *BS 1017: Part 1: 1977: Sampling of coal* (London: British Standards Institution) (1977)

9 Site survey — energy conversion and distribution

9.1 Boiler plant

In many buildings, a major proportion of purchased fuel is likely to be used for boiler plant. Attention to efficient boiler operation is therefore of fundamental importance.

The procedures described below are also largely applicable to other fired units, including warm-air heaters and thermal fluid heaters.

9.1.1 General

— Determine the type, rating and age of each boiler and burner and the types of fuel that can be used.

— Establish the annual fuel consumption of each boiler or group of boilers. This could be on a boilerhouse basis or, preferably, by service if heating and hot water boilers are separate. If not metered this might alternatively be determined from an assessment of the connected loads and the seasonal efficiency.

— Note the physical condition of each boiler and associated plant. This could well establish the life expectancy of plant items; this should have a strong effect on any recommendations made following tests.

— If possible, examine the heat transfer surfaces for signs of fouling or corrosion which could indicate the need for boiler or burner maintenance. Look for signs of condensation or low back-end temperatures in relation to the manufacturer's recommendations.

— On steam plant, if the water side and remote parts of the flue gas side are not accessible for examination during the survey, refer to the latest boiler insurance inspector's reports.

9.1.2 Operation

— Examine the method and suitability of control of boiler operational time in relation to the type of load, e.g. optimum start control for variable preheating requirements.

— Examine the control of the boiler firing rate in response to demand; check that the water temperature or steam pressure match requirements within both high and low limits. High flow temperatures may increase system losses directly and low temperatures may lead to extended operating periods if user requirements are not satisfied. Check that on/off type burners are not cycling frequently and that high/low or fully modulating burners are set to match the range of demand experienced. Boilerhouse chart recordings of output, flow temperature or flue gas temperature can be a good source of information on boiler cycling.

— Examine logbook data and question operators to ascertain the load pattern, operating practices and maintenance procedures. The need for changes in operating practice or revision of service intervals may be highlighted. An insight can also be gained into the behaviour of a system.

— For multiple-boiler installations examine the methods for controlling the number of boilers kept on-line in response to changes in load. Review the need to keep additional boilers on standby. Consider automatic sequence control where not already fitted — even a saving of 2–3% could be worthwhile.

— Review the methods of selecting boiler duty. Lead duty should normally be given to the most efficient boiler or the one best matched to the load — a smaller boiler on full load is likely to be preferable to a larger lightly loaded boiler. Note that there might be a financial advantage in operating a less efficient boiler if it can use a cheaper fuel than other plant. However, any cost savings may be made at the expense of increased emissions.

— Identify opportunities for restricting standing losses caused by flow through idle boilers, e.g. on the gas side automatic flue gas dampers or air dampers on burner inlets; on the water side facilities to prevent idle boilers from remaining at full system temperature. Complete seasonal isolation of some or all plant might be possible. Ensure that manufacturer's instructions are followed with regard to protecting the boiler and flue from corrosion.

9.1.3 Thermal efficiency

The thermal efficiency of any boiler should be determined both for the purpose of analysing energy use and with a view to improvement. Efficiency can often be improved by simple adjustment of the burner to minimise excess combustion air, other than on domestic sized or commercial gas-fired boilers with atmospheric burners. British Standard test procedures[1] should be used. Efficiency is determined either by measuring the heat output in relation to the heat input (the direct method), or by inference from measurement of the losses (the indirect method).

For short-term tests performed during a survey, when a boiler is on load, the indirect method is usually appropriate. The direct method, on the other hand, can be extended to provide information on long-term or seasonal efficiency over a range of load conditions but is normally only applied to major installations.

Table 9.1 presents the minimum instrument requirements for calculations of the thermal efficiency by the indirect method. Some instruments can measure several of the parameters and calculate the efficiency approximately.

To examine a boiler plant in detail the use of recording instruments is preferred, to confirm steady-state conditions or monitor any patterns. Where meters are permanently installed these should only be used if their calibration is reliable and any necessary correction factors can be determined.

— Take care to ensure that readings obtained at the flue outlet of a boiler are not influenced by air inleakage via inspection doors or flue joints. Inleakage will reduce the CO_2, CO and smoke readings and increase the O_2 reading, giving a false indication of combustion conditions. Certain designs of flue that dilute combustion products are unsuitable for this type of testing.

— Depending on burner type, obtain boiler test data at more than one firing rate to ensure that performance is determined over the full operational range. This may be at high and low firing rates for a two-stage burner or at maximum, medium and minimum settings if a fully modulating burner is fitted.

— Interpret the results of flue gas analysis to indicate how performance can be improved, e.g. by adjusting burner settings controlling the air-to-fuel ratios on pressure jet, blown gas or rotary-cup type burners.

For the main fuel types Table 9.2 gives typical expected values, on a dry basis, for CO_2 and O_2 contents, CO concentrations and smoke numbers in flue gas.

Table 9.1 Instrument requirements for testing boiler efficiency by indirect method

Parameter	Instrument
CO_2 or O_2 content	Portable indicators (electronic or chemical absorption)
CO concentration or smoke no.	Portable indicators (electronic or hand pump with discoloration phase indication)
Flue gas temperature	Digital electronic (with thermocouple)
Combustion air temperature	Digital electronic (with thermocouple)
Fuel firing rate	Permanent meter(s) if installed and working (or qualitative, e.g. high/low/mid-fire)

Low CO_2 contents or high O_2 contents generally suggest that excess combustion air is being used. Better results are normally expected for major boiler plant than for minor plant. For solid and gaseous fuels the CO content of the flue gas represents partially burnt fuel, while for liquid fuel, the smoke number gives an indication of combustion cleanliness.

— Calculate the heat account for each boiler using the indirect test results. Typical heat accounts for solid, liquid and gaseous fuel fired non-condensing boilers tested at full load are shown in Table 9.3. Note that for plant with variable firing rates the radiation, convection and conduction losses increase at part load, e.g. double at half-load. A simplified assessment of stack losses can be made by using the graphs in Figures 9.1–9.6[2]. These figures do not include any allowance for unburnt fuel or radiation loss.

Higher efficiencies should be achieved with large, modern plant. Lower values can be expected for small boilers of outdated design.

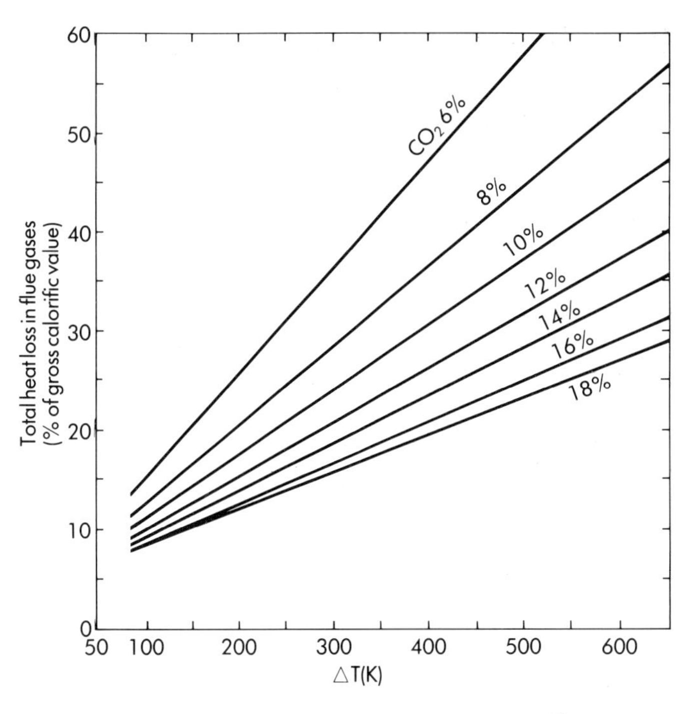

Figure 9.1 Flue gas loss — bituminous coal (rank 702)[2]. ΔT is the difference between flue gas and room temperatures.

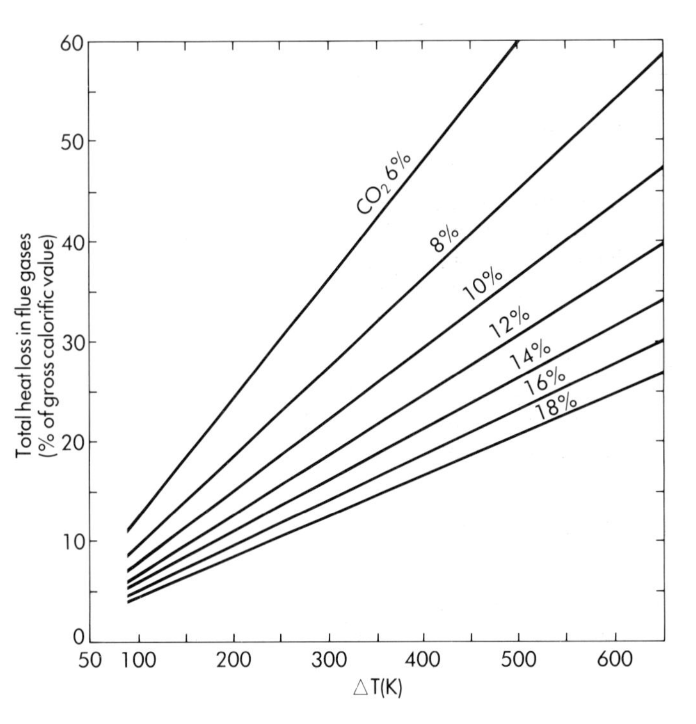

Figure 9.2 Flue gas loss — coke[2]. ΔT is the difference between flue gas and room temperatures.

Table 9.2 Typical CO_2 and O_2 contents by volume expected in flue gas (dry basis)

Fuel	Min. fire		Full fire		CO (ppm)	Smoke number (maximum)
	CO_2(%)	O_2(%)	CO_2(%)	O_2(%)		
Coal	11.0	8.5	14.0	5.0	2–500	
Fuel oils	11.5	5.5	13.5	3.0	—	0–1 (Class D) 2–3 (Classes E, F, G)
Butane	9.4	7.0	12.0	3.0	2–400	—
Propane	9.2	7.0	12.0	3.0	2–400	—
Natural Gas	8.0	7.0	10.0	3.3	2–400	—

Table 9.3 Typical analysis of losses for non-condensing boilers tested in accordance with BS 845[1]

Parameter	Fuel		
	Coal	Oil	Natural gas
Losses (%)			
Dry flue gas	12.0	10.5	9.5
Moisture loss	5.1	7.5	11.5
Unburnt gas	0.1	Neg.	0.1
Carbon in ash etc.	1.0	—	—
Radiation etc.	2.0	2.0	2.0
Total losses (%)	20.2	20.0	23.1
Thermal efficiency by difference (%) (based on gross calorific value)	79.8	80.0	76.9

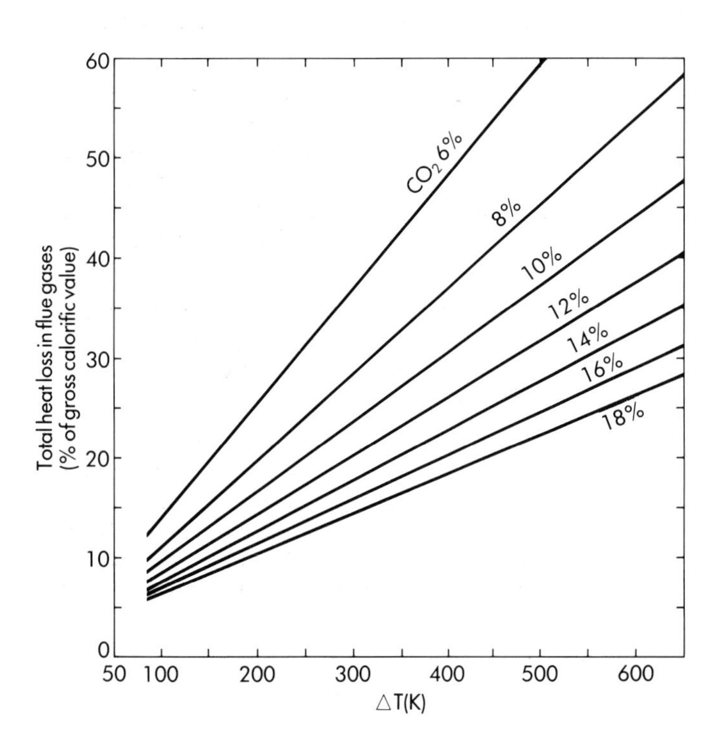

Figure 9.3 Flue gas loss — anthracite (rank 102)[2]. ΔT is the difference between flue gas and room temperatures.

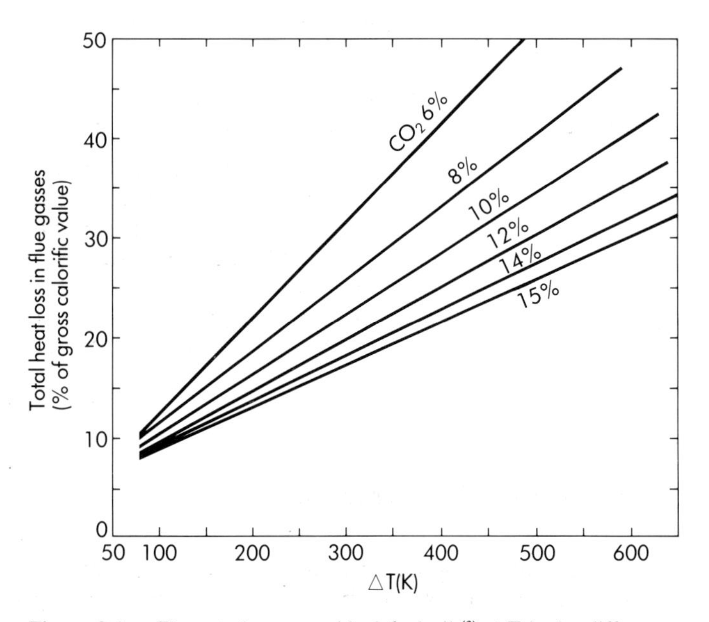

Figure 9.4 Flue gas loss — residual fuel oils[2]. ΔT is the difference between flue gas and room temperatures.

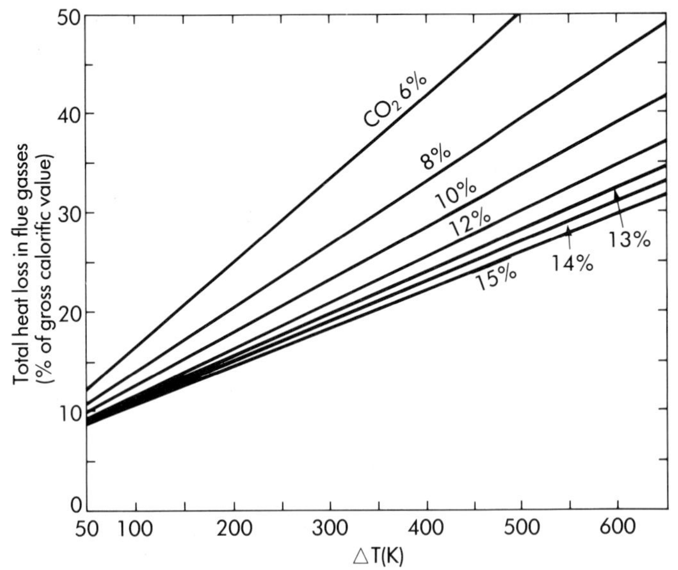

Figure 9.5 Flue gas loss — distillate fuel oils (gross calorific value 46.5 MJ/kg; specific gravity 0.79)[2]. ΔT is the difference between flue gas and room temperatures.

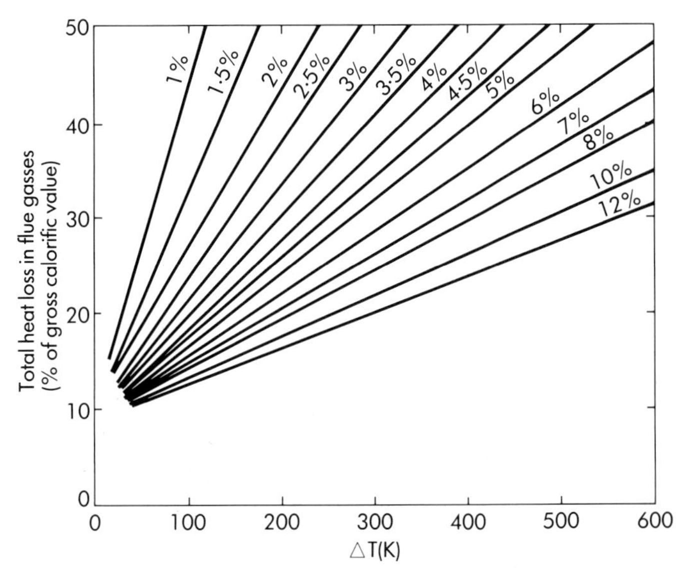

Figure 9.6 Flue gas loss — natural and manufactured gas[2]. ΔT is the difference between flue gas and room temperatures.

9.1.4 Seasonal efficiency

Over a season, the average efficiency, determined from heat output compared with heat input, provides a long-term performance measure that takes account of operational cycles.

— Where reliable records exist, calculate the seasonal efficiency directly from metered fuel input and heat output. Results should be used cautiously if there is any uncertainty in the accuracy of the metering.

Table 9.4 Typical seasonal efficiency values for package LPHW boilers up to 500 kW[3]

Boiler/system	Seasonal efficiency (%)
Condensing boilers	
Underfloor or warm-water system	90
Standard size radiators, variable temperature circuit (weather compensation)	87
Standard fixed temperature emitters (83/72°C flow and return)	85
Modern high-efficiency non-condensing boiler	80–82
Conventional boilers	
Good modern boiler design closely matched to demand	75
Typical good existing boiler	70
Typical existing oversized boiler (atmospheric cast-iron sectional)	55–65

— For basic surveys, using the measured thermal performance as a guide, assess the seasonal efficiency. Table 9.4 gives typical values[3] for seasonal efficiencies. Poor load matching and inadequate maintenance could reduce the values by up to 30% in extreme cases.

— For detailed investigations the expected seasonal efficiency can be estimated allowing for actual plant and equipment sizing and varying load conditions. CIBSE *Applications Manual AM3*[3] describes a model suitable for LTHW boilers. The model takes account of part-load efficiency and the times of operation at various loads.

Typical curves[3] for part-load efficiency are given in Figure 9.7.

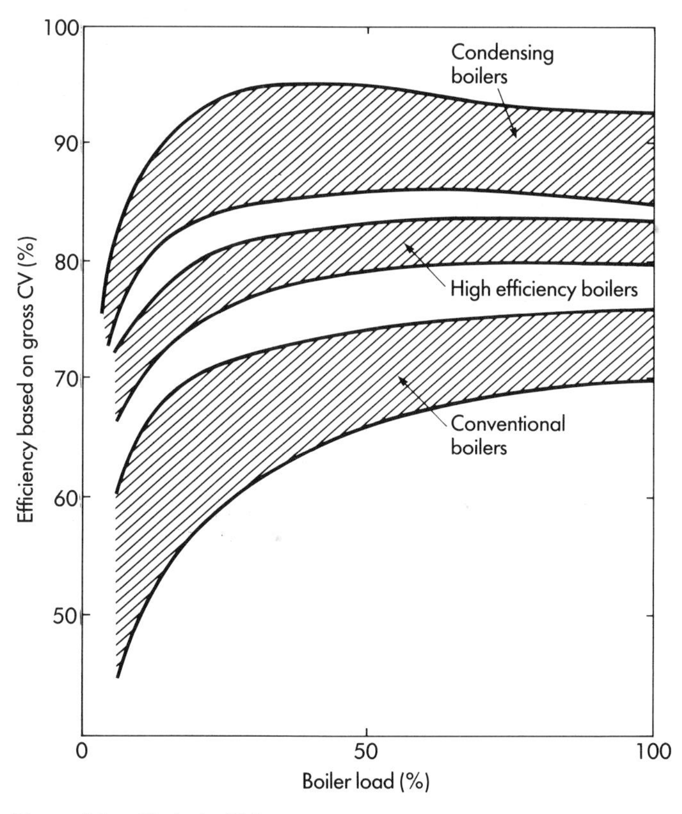

Figure 9.7 Typical efficiency ranges at part load for package LTHW boilers up to 500 kW[3]

9.1.5 Feedwater and blowdown (steam boilers only)

As water is evaporated so dissolved solids introduced with the feedwater concentrate in the boiler and settle to its base. The methods and requirements for removal of the total dissolved solids (TDS) that have accumulated should be reviewed:

— Consult the boiler manufacturer and/or water treatment specialist to establish the maximum recommended boiler water TDS value. The value will depend on the type of boiler and the operating pressure. Typical values are between 2000 ppm and 3500 ppm for most shell-type boilers operating at below 10.0 bar pressure.

— Review the arrangements for using condensate return and mains make-up as feedwater. Condensate usually has a very low TDS and it is therefore essential to ensure maximum return and re-use as boiler feedwater in preference to mains water.

— Examine the procedures for blowdown and ensure that it is kept to the minimum necessary. Water may be blown down either intermittently by manual use of the main blowdown valve, or continuously via a valve set manually or adjusted automatically. The process can be extremely wasteful of energy and water if not controlled properly.

— Calculate the amount of boiler water blowdown B as a percentage of the feed rate as follows:

$$B = \frac{100\ T_f}{T_b - T_f} \quad (\%)$$

where T_f is the TDS of feed in ppm and T_b is the TDS of boiler water in ppm.

Note that correct control of TDS saves on treatment costs as well as energy and mains water consumption.

— Investigate the opportunity for heat recovery, if not already practised, using a continuous blowdown system discharging to a flash vessel. Flash steam formed in the vessel can normally be fed directly into the boiler feed tank. Condensate containing the concentrated solids is usually discharged from the flash vessel to drain via a coil in the cold make-up tank which it can serve to preheat.

9.1.6 Flue gas heat recovery

— Investigate the opportunity for flue-gas heat recovery to improve the overall boiler thermal performance, particularly on large gas-fired plant operated for long periods.

— Consider the potential for application of modern gas-fired condensing boilers[3] which are designed to condense water vapour and thereby recover its latent heat. Considerably more sensible heat is also extracted from the flue gas in such plant.

— Review the operation of existing economisers used to recover heat. With solid fuel fired boilers the flue gas is commonly cooled to between 150°C and 160°C but below about 145°C the heat recovery equipment may be bypassed to avoid acid vapour condensation. With dual-fuel (gas and fuel oil) firing it is usual to bypass heat recovery equipment when oil firing to avoid corrosion and acid smut emission.

— Check that maintenance procedures are adequate to minimise the effect of fouling of heat exchangers when using solid or liquid fuels. Condensation on surfaces will attract solids present in the flue gas (carbon, ash etc.) and the unit may rapidly become ineffective.

9.1.7 Insulation

The outer surfaces of boilers and associated steam and hot water vessels, pipework and fittings should be insulated for two reasons. Firstly, to reduce heat loss and save energy and secondly to prevent accidental burns.

— Examine the standard, condition and suitability of insulation. Note that insulating covers for valves and fittings should be easy to remove and replace for maintenance.

— Determine the economic thicknesses for the particular circumstances. A case can often be made for upgrading or replacing insulation.

9.1.8 Stationary source emissions

Neither gaseous nor solid emissions from boiler plant are normally measured as part of an energy survey at present. However, they do reflect the quality of operation and the total emissions are directly related to the quantity of fuel used. Increasing concern for environmental matters may in future make this a subject for inclusion in detailed surveys.

9.1.9 Water treatment (steam boilers)

Mains water, although supplied as potable, contains many impurities which could be harmful or even dangerous if allowed to enter a boiler water system. Treatment is thus necessary but should be controlled at the minimum necessary level, ideally by keeping make-up water requirements to a minimum.

— Check that the appropriate equipment is installed to produce make-up water ideally of zero hardness, free from dissolved gases and insoluble solids, and alkaline, with a pH in the range 7.5 to 9.5.

— Review the routine testing and dosing procedure.

— Ensure that any criteria laid down by the water treatment specialist for the particular boiler plant are being adhered to.

9.2 Incinerators

— Determine the age and general condition of the incinerator and any manual or automatic feed mechanism.

— Establish the sources, type and quantities of waste materials available for burning. Note whether different wastes are available separately or require sorting.

— Identify any use of conventional fuel (gas or oil) in the main combustion chamber or in an after-burner to minimise smoke. Obtain details of fuel consumption.

— Determine the approximate likely heat output from the waste. Take an average calorific value based on the typical constituents of the waste. Table 9.5 gives some typical calorific values of common waste products.

Table 9.5 Typical calorific values of common types of waste

Waste type	Typical calorific value (MJ/kg, dry basis)
Plastics	37.0
Textiles	16.3
Wood (air dried)	15.8
Paper, cardboard	14.6
Vegetable matter	6.7

— If there is no heat recovery examine the feasibility of adapting or replacing existing plant. Investigate local potential heat users with a heat requirement coincident with incinerator operation. It is essential that the pattern of waste supply be matched to that of the prospective load. Storage facilities may assist in the matching of supply and demand.

9.3 Electricity generating and combined generation (CHP) plant

Electricity generating plant may range from small reciprocating engines to major steam or gas turbines. Small-scale plant[4] has the most widespread application, but similar general considerations apply in almost all cases:

— Determine the type, rating and age of plant.

— Establish the fuel types, costs and quantities used and note where alternative, cheaper fuels might be suitable.

— Assess the physical condition and review the maintenance procedures for plant and ancillaries. Ensure that servicing is adequate to maintain optimum performance.

— Establish the operational times and load patterns on a daily/weekly/seasonal basis if possible. The existence of a substantial base load for both heat and power for much of the year is a prerequisite for continuous duty. It will largely determine the economic size of plant.

— Where reliable records are available determine the overall efficiency of the plant by relating useful energy output to energy input. It may be possible to estimate either input or output if only the other is known, and an efficiency is assumed for the type of plant. Some typical examples are given in Table 9.6 for guidance only. In practice, the particular plant arrangements may produce a different balance.

Table 9.6 Examples of energy balances for CHP plant as a proportion of heat input (%)

Plant	Electrical power output	Heat available for recovery	Other losses
Small gas engine	25–30	55–60	10–20
Diesel set	38	57	7
Gas turbine	18	74	8
Steam turbine	9	83	8

In connection with Table 9.6, note that the heat available from gas and diesel engines includes a contribution from the cooling circuits. Further low-grade heat may also be recovered from these circuits.

— Review plant performance over a range of loads. Comparison with manufacturer's performance curves or previous tests results may be possible. Full performance testing can be a major task for large-scale plant but could be justified in a comprehensive survey.

— Calculate the cost of electricity, heat or steam produced and compare with other sources, e.g. boiler plant and imported electricity. The comparison must relate to different purchase tariffs that would apply for day/night or seasonal loads. Internal accounting procedures will determine the maintenance and operating costs, additional to that of fuel, included in the calculation.

— Confirm the economic operating times, e.g. continuous, day only, winter only, by identifying two or three possible scenarios and evaluating them. The results will be determined largely by the cost or value of energy imported, consumed or exported but other costs must also be considered. Small engines might require only simple routine maintenance while for major, attended plant there is also a significant manpower cost.

— Investigate opportunities for making greater use of available heat. In addition to high-grade heat in the exhaust gas of an engine, lower-grade heat may be available from oil cooling, or the engine cooling water.

— Consider increased use of plant provided for peak-lopping or standby purposes. Note that plant may not be designed for continuous duty — the manufacturer should be consulted.

— Consider using the engine to drive an air compressor or pump, incorporating heat recovery.

9.4 Refrigeration plant

It is assumed here that plant is associated directly with air conditioning systems although the survey requirements could apply equally to other cooling applications. Refrigeration can represent a major electrical load with a substantial cost in energy, although this is often outweighed by associated fan and pumping costs. There may be many opportunities for savings, some of which can be identified without specialist knowledge of refrigeration.

9.4.1 General

— Identify the type of plant and its rating. Chilled water is normally produced by means of mechanical vapour compression using electrically driven reciprocating, centrifugal or screw compressors. The alternative absorption process requires heat input, usually from hot water or steam. Various wet or dry systems may be used to cool condenser water. Modern packaged gas absorption units are air cooled and do not require cooling towers.

— Note the age and condition of plant. If replacement is due then alternative systems should be considered.

— Establish the annual energy consumption of the plant, normally electricity. Compressors, cooling and condenser water pumps and cooling fans must all be included. Condenser water costs may also have to be added where water is dumped rather than recirculated. When assessing electricity costs remember that for summer operation the maximum demand charges may be relatively small. However peak summer loads can determine the overall supply capacity that is likely to incur a fixed charge throughout the year.

— Determine the consumption of unmetered, fixed-load plant from measurements of load and running hours. In an air-conditioning application a centrifugal compressor typically requires a power input in the region of 0.2 kW per kW cooling. A reciprocating compressor requires slightly more and a screw compressor slightly less. An absorption machine could require perhaps seven times this rate of energy input for the same cooling duty. However, it might usefully reduce electrical demand charges and offer CFC-free operation.

— Variable-load plant will normally require direct measurement of consumption. Total refrigeration system efficiency will normally fall as the load reduces. At 25% of full load the power requirement could still be 40 to 50% of that at full load.

— Establish the operational times and patterns of use of plant and equipment. Confirm when plant is on-line and cooling is made available.

— Examine the demand for cooling and review the requirement for refrigeration. Cooling towers alone may sometimes cool water sufficiently to allow refrigeration plant to be by-passed other than on hot summer days. Uncooled ambient or recirculated air could be acceptable except when peak internal gains occur. Dehumidification using cooling coils is usually not needed until relative humidity reaches typically 60% for computer rooms or 70% for general occupancy. Substantial savings can be made if chilled water pumps are switched off when the cooling load falls to a level where cooling is not entirely necessary.

— Identify when plant can be shut down completely, e.g. during winter, overnight or at weekends. Alternatively, consider potential for pre-cooling a building, or storing cooling capacity in ice formed in purpose-made storage vessels, using cheaper off-peak electricity.

— Investigate opportunities for heat recovery in the form of warm air or hot water. Suitable heat demands should be identified and matched to the quality and availability of recovered heat, e.g. cold make-up water feeding an adjacent hot water service might require preheating to between, say, 40°C and 60°C. Heat available for reclaim is typically equivalent to slightly less than the power absorbed by the compressor in a vapour compression system. Existing heat recovery systems should also be reviewed to ensure economic operation.

9.4.2 Chiller plant

— Measure water inlet and outlet temperatures and relate these to design data. Chilled water flow temperatures of between 5°C and 9°C are expected for most systems. With temperature differences of perhaps only 3°C it is important to be able to measure temperature to a resolution of ±0.1°C. This may not be possible with some standard measuring equipment — platinum resistance or mercury-in-glass thermometers could be required.

— Note compressor suction and discharge pressures where gauges are fitted and compare results with design specification. Refer if necessary to the manufacturer's data or refrigerant pressure — enthalpy diagrams. Approximate pressure data for some common refrigerants are given in Table 9.7.

Table 9.7 Appropriate compressor suction and discharge pressures when evaporating at 5°C and condensing at 40°C

Refrigerant	Suction pressure (bar)	Discharge pressure (bar)
R11	0.50	1.75
R12	3.63	9.61
R22	5.84	15.34
R502	6.78	16.77
R717	5.16	15.55

Low suction pressures can indicate poor flow or lack of refrigerant. Faulty compressor valves can cause low discharge pressure. High discharge pressures may result from various faults including a fouled condenser or faulty cooling tower.

Operation with pressure ratios or temperatures differing from the original design may result in significant loss of efficiency.

— Investigate the facilities for capacity control for load matching. If several machines operate in parallel check that there are controls to avoid part-load operation and keep the minimum number of machines running on full load. Individual pressure controls can be set to allow compressors to cycle in sequence. Centrifugal and screw compressors in particular have poor part-load performance. Variable-speed drives can be considered for reciprocating compressors — operation at half-speed can increase compression efficiency by 10%. This could replace stop/start or cylinder unloading control. Systems which control capacity by throttling gas flow to the compressor or allow gas to by-pass the condenser are extremely inefficient and should generally be avoided.

— Check maintenance procedures, including cleaning of heat transfer surfaces and water treatment. Common faults, notably fouling and faulty valve operation, can reduce performance progressively. This both restricts the available duty and increases running costs, perhaps by as much as 50% in some cases. A reported increase in oil consumption can usually be taken as a sure sign that fouling will be a growing problem.

9.4.3 Condensers

— Check that water temperatures/refrigeration pressures/air temperatures are correct and that performance is satisfactory. A water temperature rise of approximately 5°C is typical across a shell and tube condenser, with a leaving water temperature approaching 5°C below the condensing temperature. Air cooled condensers are usually designed to provide a condensing temperature in the range 10–20°C above entering air temperature.

— Check that fan and pump controls are operating correctly. Sequence control should also be applied for load matching. Some split system and packaged units are suitable for head pressure control to minimise condenser fan operation in winter.

— Check maintenance procedures, including cleaning of heat transfer surfaces and screens/filters, water treatment and refrigerant leaks. A build-up of air and non-condensable gases can further increase condensing temperatures and reduce efficiency where maintenance is neglected.

9.4.4 Cooling towers

— Check tower performance by measuring inlet and outlet temperatures and ambient wet-bulb temperature. Relate these to specified requirements. The leaving water temperature should typically approach 5°C above the wet bulb temperature of the entering air.

— Check for correct time, temperature and sequence control of fans and thermostatic control of immersion heaters.

— Check maintenance procedures, including cleaning of screens and strainers, and water treatment, all vital to avoid bacterial growth. Poor sprays and damaged or missing baffles left unreplaced will reduce cooling efficiency and lead to longer plant running times.

— Investigate the balance of water flow through the condenser by-pass, if fitted. Too little flow through the condenser can raise the condensing pressure. Restriction of flow through the bypass will increase the pumping head. It may be possible to check the bypass regulating valve setting against commissioning data as a first step.

9.5 Distribution systems

The efficiency of distribution systems becomes significant for large buildings or sites and directly affects the overall efficiency of end use. Under some circumstances it may even be found that central plant is running purely to maintain the distribution losses. On extensive sites the distribution losses of central systems need to be distinguished from the end use.

9.5.1 Electricity

Sites metered at high voltage have their own transformers, all of which contribute to standing losses.

— Examine the loadings of all high/low voltage transformers with a view to taking very lightly loaded transformers out of service permanently, provided that their loads can be transferred and accommodated elsewhere.

The efficiency of a large, fully loaded transformer should be over 99%. For a small, lightly loaded transformer the efficiency may be only 97%. Only sites which have undergone substantial alterations from the original design are likely to present good opportunities. The need for spare capacity and/or security of supply must also be taken into account. Note that a fully isolated transformer can become unserviceable due to the ingress of moisture to the transformer oil.

— Where power is distributed via sub-stations the load at each should be examined in the same way as that for the complete site.

9.5.2 Piped services

The main services of interest are hot water, chilled water, steam and compressed air, all of which consume energy in their generation and distribution.

Direct losses

— Inspect systems visually for leaks or signs that they are occurring. Flooded ducts, plumes of steam and corrosion are some of the obvious indications.

— Listen to compressed air and steam systems for evidence of leaks, preferably during quiet periods, paying particular attention to any fittings such as flexible couplings and drain points which may have been left open.

— Identify opportunities for isolating equipment with a live end loss, such as pneumatic controls, when not in use. Local solenoid valves, linked to operation of equipment, can sometimes be fitted in compressed air lines to control the supply automatically.

— Determine and account for any imbalance between supply and end use as indicated by discrepancies in metered quantities or assessed consumptions. Unaccounted consumption may be due to leakage.

— Observe operation of plant and apparent load under conditions of little or no demand. If consumption appears to remain high even when plant is switched off then system losses are a strong possibility.

— Evaluate the losses identified by direct measurement or assessment. Leaking water can sometimes be collected and the volume measured over a known period. Compressed air leaks can be assessed individually if the pressure and size of hole are known or collectively by monitoring pressure decay or plant running time with all end users isolated.

Heat losses/gains

In addition to the direct heat loss associated with a leak, heat transfer will occur between a pipe and its surroundings.

— Examine the type, condition and thickness of insulation on pipes and fittings. Note where additional weatherproofing or other protection is needed.

— Determine the economic thickness[5] for the particular application and consider upgrading. Some examples of heat loss with varying insulation thicknesses, using preformed rigid fibrous sections (which include rock and glass fibres), are given in Figures 9.8 and 9.9[5]. The heat loss from an uninsulated valve is equivalent to that from 1 m of uninsulated pipe.

— Identify the quantity, type and location of missing, wet or damaged sections of insulation that should be replaced.

Pipework utilisation

— Service mains should be surveyed to confirm that all end users are correctly identified. The potential for

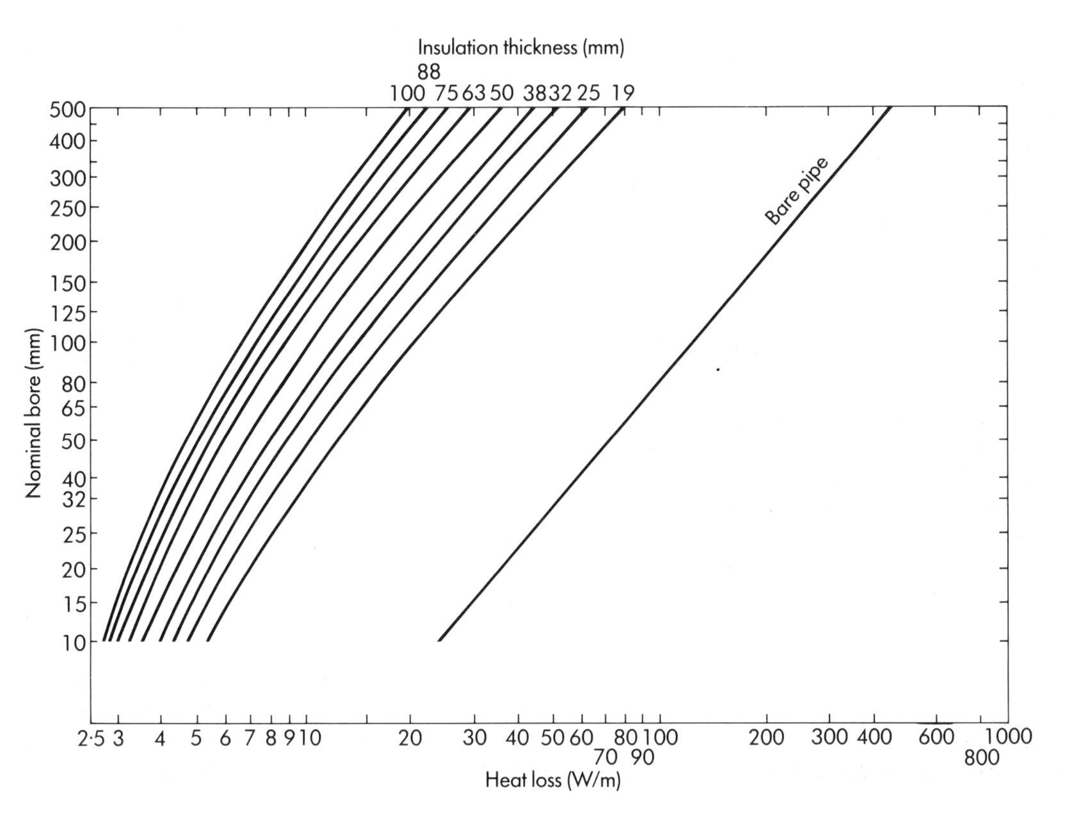

Figure 9.8 Heat loss for pipes with surface temperature of 50°C with varying insulation thickness[5]

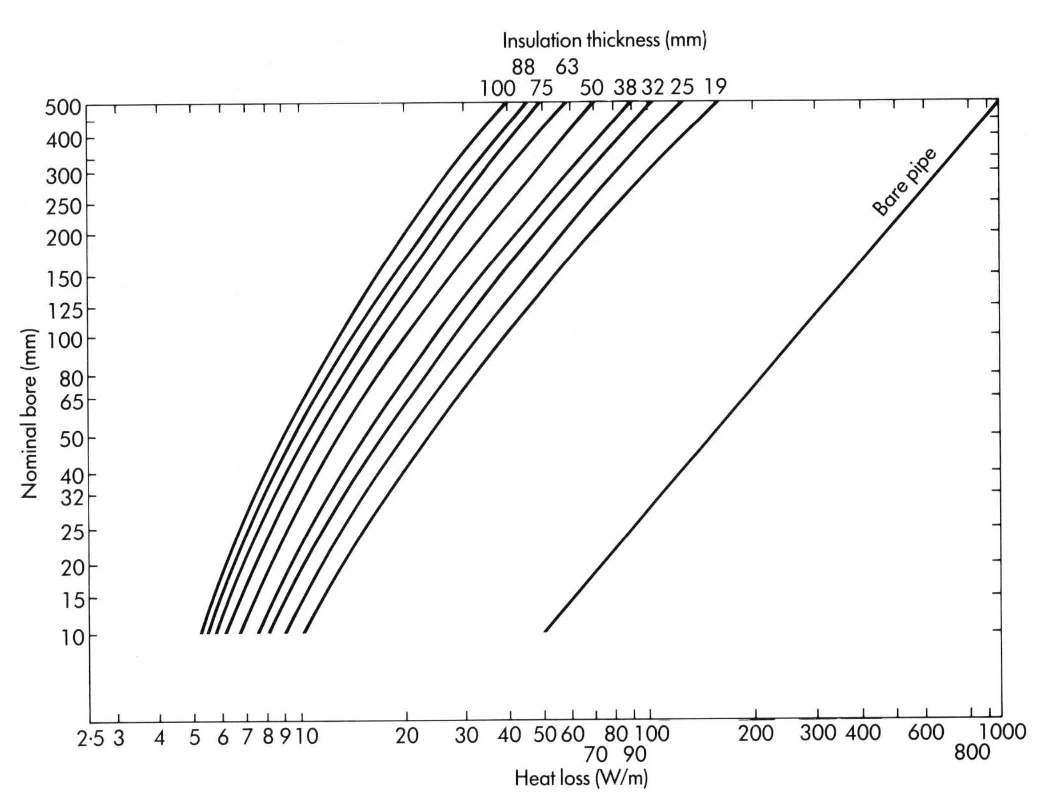

Figure 9.9 Heat loss for pipes with surface temperatures of 75°C with varying insulation thickness[5]

rationalisation should be examined where mains are under- utilised. When a main is grossly oversized and in poor condition it may be possible to justify early replacement. Reconnection of an end user to an alternative, convenient source of supply may even allow an inefficient main to be decommissioned.

— Identify sections of the service which are redundant but remain live. All redundant services should be capped, if not removed altogether.

— Identify opportunities to reduce pumping rates in hot water systems, having due regard to system design and required temperature drops.

9.5.3 Steam systems

Distribution pressure

— Consider using lower distribution pressures to reduce losses. The minimum permissible pressure will be determined by the pressure/temperature needs of the end users and the capacity of the pipework to handle increased steam volume.

Steam traps

— Check that traps are not allowing steam to pass. Portable meters can be used for test purposes, but the installation of permanent in-line devices should be considered for routine checks, particularly where access is difficult.

— Check that bypasses have not been left open. Sometimes a bypass is opened to overcome problems such as air-locking which need further investigation.

Specialist advice should be sought on correct trap application where faults occur persistently.

Condensate return

Maximum return of condensate for use as boiler feed is essential to recover heat and to reduce the need for make-up water and treatment.

— Assess the amount of condensate returned to the boiler-house on steam plants. This can usually be established by the difference between readings of the boiler feed-water meter(s) and the make-up water meter(s). Where a condensate meter is installed it is likely to read 'fast' because, operating volumetrically, it will not differentiate between steam and condensate. Consider the quantity returned as a proportion of the mass of steam generated. Condensate should be available from virtually all steam users, except where steam is injected live or where condensate may be contaminated.

— Identify condensate losses due to leaks or overflow to waste from condensate collection vessels.

— Note steam users from which condensate is not returned, possibly due to the nature of the process or its location, and consider the potential for recovery.

Flash steam

— Identify problems of flash steam formation, commonly found where condensate is discharged to an open-vented vessel. This can be caused by faulty steam trap operation.

— Consider ways of eliminating flash steam loss. A flash vessel could be installed to produce low-pressure steam. Alternatively it may be possible to precool pressurised condensate and recover the heat, for instance for space heating.

9.5.4 Decentralisation of boiler plant

The full economic implications of decentralisation, or indeed centralisation, need separate specific study. It should, however, be possible to evaluate the likely potential within the scope of a survey.

Where plant is centralised the cyclic/seasonal pattern of heat use should be examined. The overall distribution loss as a proportion of the heat input should be assessed to indicate the efficiency of the distribution system. The loss may be equivalent to only 0.5% of the boiler fuel used on a large site to meet the peak winter load when all heating plant is in use.

In summer, the distribution loss may be reduced but it is still likely to represent a far higher proportion of the overall load, perhaps over 50%. The provision of alternative local plant could then be worthwhile.

A large site where the end users have widely diverse load patterns could present a good case for decentralisation. There are probably occasions when few, if any, of the users are demanding heat but the entire distribution system is kept live. Selective installation of local plant with individual user control might be justified. Full consideration must, however, be given to the possible technical constraints, not the least of which is space availability.

Particular thought must be given to fuel policy. Distributed plant may be limited to a single fuel type e.g. gas. With central plant, alternative fuels or advantageous contract prices might be available. Furthermore, changes of fuel can usually be accommodated more readily in response to changes in market prices when only central plant has to be converted. The lowest overall operating cost, including fuel and maintenance, is not necessarily achieved through the most energy-efficient scheme.Consideration might also be given to possible future combined heat and power schemes where a centralised system could offer a useful base load.

9.6 Summary of section 9

- Boiler plant often accounts for a major proportion of purchased fuel. Thermal performance should be assessed.
- With many boilers savings can often be made by simple adjustment of burner settings.
- Use of waste materials as an energy source requires the availability of waste to be matched with a suitable heat load.
- The economics of on-site electricity generation may need special study.
- Refrigeration can represent a major electrical load and many opportunities for savings might be identified even with limited specialist knowledge.
- Distribution systems can significantly affect overall system efficiency. Mains and standing losses can account for the majority of boilerhouse heat output under light load conditions.

References for section 9

1 *BS 845: 1987: Methods of assessing thermal performance of boilers for steam, hot water and high temperature fluids:* Part 1 *Concise procedure*, Part 2 *Comprehensive procedure* (London: British Standards Institution) (1987)

2 Fuels and combustion *CIBSE Guide* Section C5 (London: Chartered Institution of Building Services Engineers) (1976)

3 *Condensing Boilers* CIBSE Applications Manual **AM3** (London: Chartered Institution of Building Services Engineers) (1989)

4 *Guidance notes for the implementation of small-scale packaged combined heat and power* EEO Good Practice Guide No. **1** (Energy Technology Support Unit) (1989)

5 *The economic thickness of insulation for hot pipes* Fuel Efficiency Booklet **8** (London: Energy Efficiency Office) (1982)

10 Site survey — energy use

The major part of most surveys is concerned with the way that energy is used, or wasted, in meeting operational requirements. A survey does not necessarily cover all energy uses, but the principal services are usually included.

Table 10.1 System data sheet summarising services in a building

Building name/number		:
Heating	– Boiler plant type/age	:
	– Boiler fuel	:
	– Central heat source	:
	– primary	:
	– secondary	:
	– Other (specify)	:
	– Types of emitter	:
	– Zoning arrangements	:
	– Control methods	:
Air conditioning and ventilation	– Chiller plant	:
	– Humidification plant	:
	– Heat recovery	:
	– Control methods	:
Hot water system	– Boiler plant type/size	:
	– Boiler fuel	:
	– Central heat source	:
	– Storage calorifiers	:
	– Local heaters (gas/electric/other)	:
	– Control methods	:
Power and lighting	– Lamp types and % of area served	:
	– Major equipment	:
	– Lighting control methods	:
Process plant	– Give details of major plant and uses:	
Metering	– Specify metered services and users	:
Building energy management system	– Type/extent	:
	– Functions	:

When dealing with a number of distinct buildings or areas, summaries of the relevant services and systems in each building can be useful for reference. Data can be set out in the form of Table 10.1.

It is not intended that the procedures described for different services be carried out in any strict order. The aim is rather to encourage a thorough but selective approach suited to the particular combination of systems encountered.

Procedures described for particular services or equipment should be adapted for application to services or types of equipment not specifically mentioned.

10.1 Audit of energy use

A site survey allows deeper analysis than is possible in a preliminary audit.

10.1.1 Energy flows

An understanding of the energy flow through a site or building is a fundamental requirement. At site or building level it is normal to consider complete services, taking account of how they interact, and identifying useful output and losses.

The concept of energy flow is best illustrated by means of a Sankey diagram, as in Figure 10.1, representing either the relative cost or consumption of the principal energy uses. Understanding of energy flow is likely to be improved in the course of constructing a diagram of this type as data for different services are collated. The information needed for the diagram must be acquired during the course of the survey.

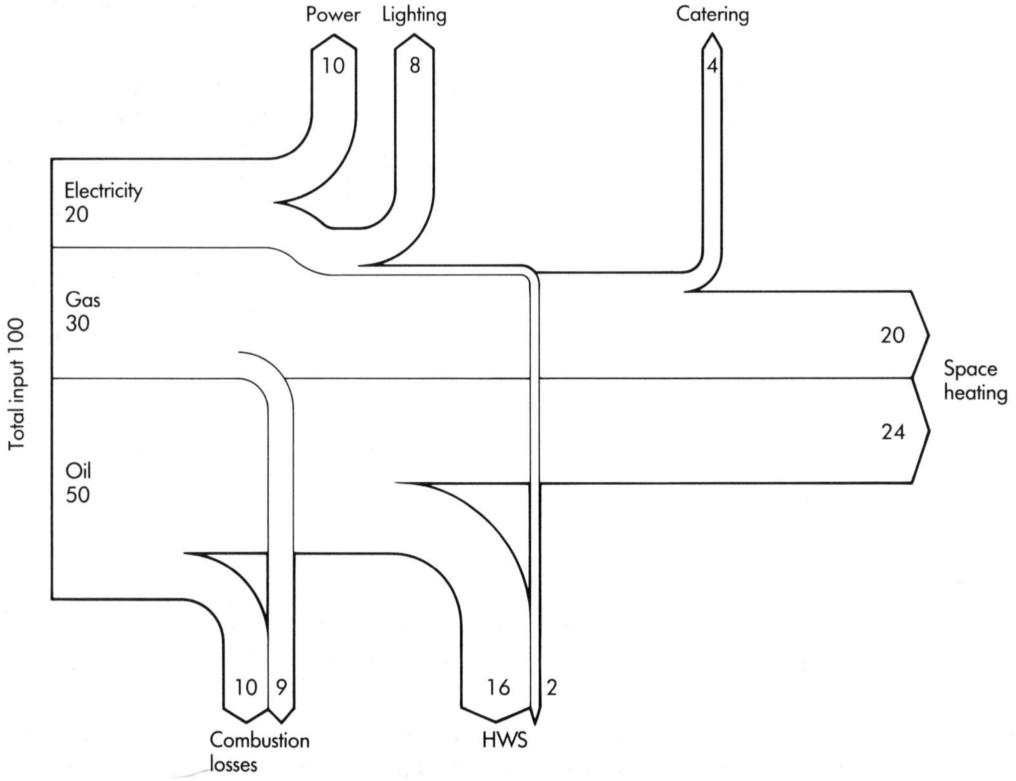

Figure 10.1 Sankey diagram showing annual gross energy consumption

The approach can be extended to individual items of equipment or processes. Initially a simple line diagram indicating the routes of energy flow into and out of a system can be prepared. Figure 10.2 is an example for a space heating system. The significance of each route of energy flow might then be evaluated, remembering that overall the energy flow across the system boundaries must balance.

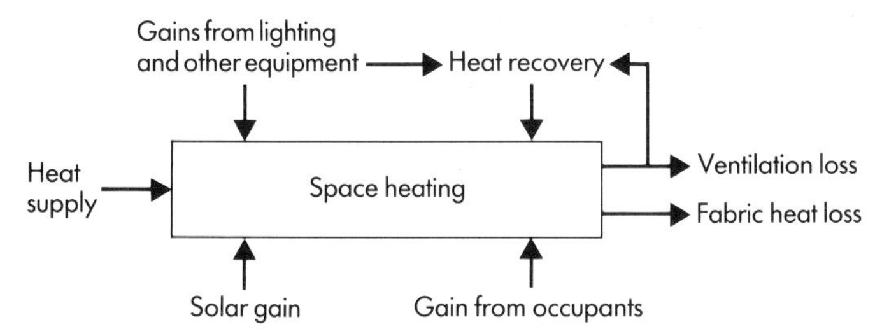

Figure 10.2 Energy flow through a space heating system

10.1.2 Breakdown of energy use

As part of a site survey it should be possible to produce a breakdown that corresponds reasonably well with real energy use.

(*a*) Prepare a table, similar to Table 5.9 in the preliminary audit, and insert all metered consumptions known from existing records as in Table 10.2. This might sometimes be extended to show distribution or combustion losses separately if desired.

Table 10.2 Summary of metered energy use

Category of use	Energy source		
	Electricity	Gas	Oil
Space heating	U	9000^{M}	2000^{M}
Hot water services	U		
Air conditioning	U	—	—
Lighting	U	—	—
Process — heating	—	500^{D}	—
— power	300^{M}	—	—
Not allocated	2300^{D}	Nil	Nil
Purchased totals	2600^{M}	9500^{M}	2000^{M}

M Metered figure
D By difference, e.g. 9500 – 9000 = 500
U Unmetered

(*b*) Compare sub-metered totals with the overall total quantities of fuel known to have been used. Account for any discrepancies due to metering errors or metering of the same use more than once at different stages. Loss or theft of fuel might also sometimes occur.

(*c*) If existing records are limited an initial breakdown may be based on the examples given in section 5 for preliminary audit. As more of the uses are evaluated, so the remainder can be adjusted until finally a balance is achieved.

(*d*) Where one meter serves a large area, or where there is no sub-metering, then sometimes energy use can be broken down further by estimating the annual consumption of the various users. Two methods most applicable to fixed loads are:

(i) By multiplying the design energy consumption of equipment of a known rating by its annual operating time, e.g. 12 kW × 8 hours/day × 5 days/week × 30 weeks/year = 14 400 kWh/year. Ratings should be measured if possible as nominal ratings or design loads may not apply to the actual operating conditions.

(ii) By multiplying the actual energy consumption of the equipment measured over a limited period by its annual operating time. Electrical loads can often be assessed for this purpose from a simple measurement of current.

(*e*) A service can be apportioned approximately on a floor area basis if a breakdown is required by buildings or areas. Further judgement must be applied where occupancy type or the level of service provided differ significantly in each area.

(*f*) Once all unmetered uses included in the required breakdown have been assessed, Table 10.2 can be redrawn as Table 10.3.

Table 10.3 Allocation of purchased energy (GJ/annum) to end users

Energy use (GJ/annum)	Energy source		
	Electricity	Gas	Oil
Space heating	100^{A}	8000^{A}	2000^{M}
Hot water services	200^{A}	1000^{A}	—
Air conditioning	1200^{A}	—	—
Lighting	900^{A}	—	—
Process — heating	—	500^{D}	—
— power	300^{M}	—	—
Calculated totals	2700	9500	2000
Purchased	2600^{M}	9500^{M}	2000^{M}
Unaccounted	(100)	Nil	Nil

M Metered
D By difference
A Assessed

The assessed figures in the example are seen to agree closely with the purchased totals. If the calculated and purchased totals disagree by more than say 5% then the assessments must be revised until a satisfactory balance is achieved. Remember that a correct total does not necessarily indicate that all assessments are accurate.

(*g*) The breakdown should finally be converted to a cost basis by applying the relevant fuel price. For most fuels this price will be the overall figure as recorded in Table 5.2. In most cases the overall price for electricity can also be used. However, if the use under assessment is quite clearly being charged for at either a day or night rate if applicable, then the appropriate unit cost should be used.

10.1.3 Estimating the energy use of some principal services

Space heating

The energy used for space heating can be estimated using a variety of techniques some of which are described in *CIBSE Guide* Volume B[1] and CIBSE *Building Energy Code* Part 2[2].

Estimates should take into account weather, building response and thermal properties, system characteristics and hours of use. In very simple terms, consumption might be expressed as:

$$F = 0.0036\,HE \times 100/\eta$$

where F is the annual fuel consumption (GJ), H is the calculated building heat loss at design conditions (kW), E is the equivalent hours full load operation (h/year), η is the seasonal efficiency of the system (%).

Calculation of the building heat loss H requires a knowledge of the thermal transmittance (U-value) of the building elements and the ventilation rate as a minimum. Reference should be made to *CIBSE Guide* Volume B[1] for a formal derivation of the equivalent hours full-load operation E. Some typical ranges are given in Table 10.4.

Table 10.4 Typical ranges of equivalent hours full-load operation

Building use	Hours
Offices, schools, lightweight buildings occupied 8 h/day, 5 day/week	Under 1500
Public retail, catering; heavyweight buildings occupied 12 h/day, 6 days/week	1500–2500
Hospitals, hotels, sheltered accommodation; continuously heated	2500–3500

Table 10.5 gives some typical examples[1] of the seasonal efficiencies of different heating systems.A more rigorous treatment will also take account of solar and internal gains and the non-steady-state conditions encountered in practice. Judgment must be applied in deciding the appropriate degree of accuracy in calculating space heating loads.

Hot water services

If the water consumption for hot water services is known, e.g. cold feed is metered, an estimate can be made of the energy consumption (Table 10.6).

Table 10.7 gives the typical efficiency[1] of some different systems.

A formula is suggested in CIBSE *Building Energy Code* Part 4[3] for use where water usage is not known:

$$E_{HWS} = 0.0864\,q_{HWS}\,A_f\,N_W \times 10^{-3}$$

where E_{HWS} is the energy consumed in providing hot water (GJ), q_{HWS} is the mean power requirement from Table 10.8 (W/m^2), A_f is the floor area under consideration (m^2) and N_w is the number of working days.

Table 10.5 Seasonal efficiencies of heating systems[1]

Type of system	Heat conversion efficiency (%)	Utilisation efficiency (%)	Seasonal efficiency of system (%)
Intermittent			
Automatic centrally fired radiator or convector systems	65	97	63
Automatic centrally fired warm air ventilation systems	65	93	60
Fan-assisted electric off-peak heaters	100	90	90
Direct electric (non-storage) floor and ceiling systems	100	95	95
† District heating/warm air systems	75	90	67.5
Continuous			
Automatic centrally fired radiator or convector systems	70	100	70
Automatic centrally fired warm air ventilation systems	70	100	70
Electric storage radiator systems	100	75	75
Electric floor storage systems	100	70	70
Direct electric floor and ceiling systems	100	95	95
† District heating/radiator systems	75	100	75

Notes

1 Very high efficiency heat generators may raise the heat conversion efficiency.

2 Solid fuel appliances working in conjuction with intermittent systems may require allowance for rekindling.

3 Heavier liquid fuels will require preheating allowance.

† Allowance should be made separately for mains heat losses on a seasonal basis.

Table 10.6 Example calculation of the annual energy requirement to heat a known quantity of water from cold to full service temperature

Annual water consumption	120 000 litres
HWS temperature	60°C
Cold make-up temperature	10°C
Seasonal efficiency of system	60%
Specific heat content of water	4.2×10^{-6} GJ/litre

Annual energy required $= 120\,000 \times (60 - 10) \times (4.2 \times 10^{-6}) \times \frac{100}{60} = 42$ GJ

Table 10.7 Seasonal efficiencies of water heating systems[1]

Type of system	Heat conversion efficiency (%)	Utilisation efficiency (%)	Seasonal efficiency of system (%)
‡ Gas circulator/storage cylinder	65	80	52
‡ Gas and oil fired boiler/storage cylinder	70	80	56
Off-peak electric storage with cylinder and immersion heater	100	80	80
Instantaneous gas multi-point heater	65	95	62
†‡ District heating with local calorifiers	75	80	60
†‡ District heating with central calorifiers and distribution	75	75	56

Notes

1 Very high efficiency heat generators may raise the heat conversion efficiency.
2 Solid fuel appliances working in conjuction with intermittent systems may require allowance for rekindling.
3 Heavier liquid fuels will require preheating allowance.
† Allowance should be made separately for mains heat losses on a seasonal basis.
‡ Dependent on the size of the heat generator, the summer conversion efficiency may deteriorate, thereby reducing the overall seasonal efficiency significantly.

Table 10.8 Mean power requirement of HWS

Building type	q_{HWS} (W/m²)
Office (5-day)	2.0
Office (6-day)	2.0
Shop (5-day)	0.5
Shop (6-day)	1.0
Factories	
5-day: single shift	9.0
6-day: single shift	11.0
7-day: multiple shift	12.0
Warehouses	1.0
Residential	17.5
Hotels	8.0
Hospitals	29.0
Education	2.0

Lighting

The installed load (kW) can be multiplied by hours in use to give consumption (kWh). The load can be determined by counting the number of fittings and identifying their indicated rating. Allowance must be made for control gear losses, except for tungsten lighting, to give the total circuit load (see Table 10.14). Between 10 and 20 W/m² is typical for fluorescent lighting when the load is related to the floor area served. Mains distribution losses associated with any individual load can be discounted as a minor factor for estimating purposes.

10.2 Space heating

10.2.1 Degree-day analysis

A graph of monthly space heating energy consumption or total boiler fuel consumption against degree days may already have been plotted at the preliminary audit stage. The data should now be studied more closely in order to account for the position of the points relative to the line of best fit. In the example of Figure 10.3, in summer, when no space heating is provided, the consumption for other uses is independent of degree days. The outlying point observed in winter, when the number of degree days is highest, might be explained by a shutdown or plant breakdown.The principal features of the graph to be recognised are:

(*a*) The intercept, theoretically representing the energy required to maintain the standing losses or other base loads. In practice, space heating is discontinued before the number of degree days falls to zero and the true base load is higher than the intercept.

(*b*) The slope, showing how the demand related to space heating varies with degree days.

(*c*) The scatter of points, reflecting the accuracy of control. A wider scatter suggests that control is not accurate, but can also be caused by a fluctuating base load. Good correlation could well result from consistent, 'accurate' control. However, it should not then be assumed that control settings need no further adjustment as heating could have been maintained consistently at the wrong level.

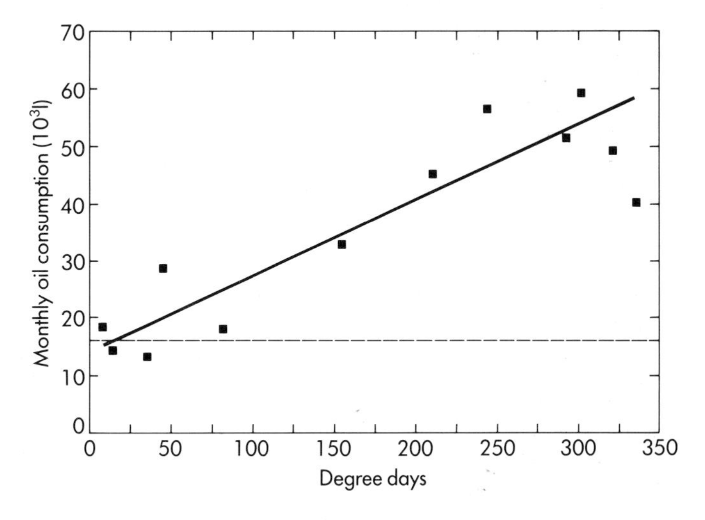

Figure 10.3 Plot of energy consumption against degree–days showing base load

10.2.2 Low-temperature hot water system

— Identify the system type and components.

— Establish the methods of control and operation, noting the facilities for automatic control and any functions performed manually.

— Check control settings, compare with the requirements for occupancy and note whether the set values are achieved. Do not assume that settings indicate the actual conditions maintained. Poorly located or inaccurate temperature sensors may give unexpected results. Timeswitches may be running fast or slow by hours or even days, even though the set programs are otherwise correct.

— Check that the equipment responds satisfactorily to controls, taking care that any tests will not seriously inconvenience the occupants or adversely affect plant. Change thermostat settings temporarily and check that equipment responds correctly. Similarly, operate overrides on timeswitches and other controls where it is safe to do so. Check that motorised valves are able to operate and that actuators have not become seized or disconnected.

— In systems where flow temperature is reduced during milder weather, the compensated flow temperature should be checked against the scheduled value for the prevailing outside air temperature. In the example shown in Figure 10.4, when the outside air temperature is 6°C a flow temperature of 63°C is expected. The case for complete system shutdown when the outside temperature reaches a pre-set level can be considered.

— Investigate zoning arrangements and local control facilities where areas on a common circuit have different needs, due for instance to building aspect, internal gains, or differences in occupancy. Discussion with occupants can help to identify poor balancing between different zones.

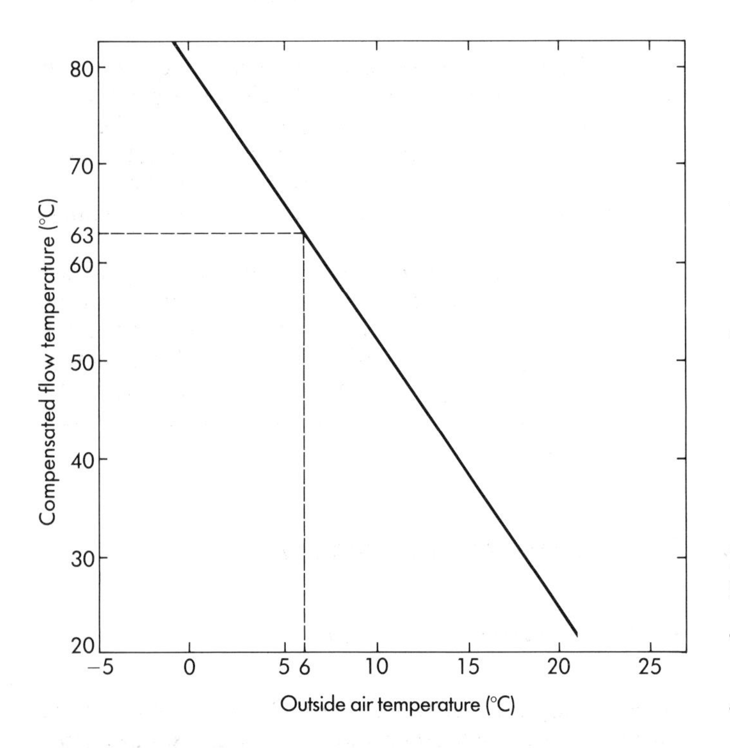

Figure 10.4 Example of compensated flow temperature schedule

— Check the cleanliness of heat transfer surfaces and unit casings and note any obstruction of air movement or heat output from emitters. Identify problems of poor water circulation or inadequate air venting. Restricted heat output tends to result in excessive operating periods and the use of supplementary heating.

— Consider applying reflective foil to the inner surfaces of outside walls behind which emitters are installed. Heat losses through the building fabric can be reduced significantly where radiators are installed at windows.

— Review the effect of direct heat output from the distribution pipework. Unlagged pipework in a roof space or void may provide no useful heat input to the occupied space and could be susceptible to freezing when not in use. Large-diameter pipework may cause excessive heat input to an occupied space if flow temperature is not controlled, even if the individual emitters it serves are all switched off.

10.2.3 Warm air units

Units are either fired individually or fitted with steam, hot water or electric heater batteries.

— Check time and room temperature controls for correct location, setting and operation.

— Inspect the condition of fired units and check for correct burner and fan operation, cleanliness of heat transfer surfaces and filters, and integrity of casing and insulation. Test the thermal efficiency as described for boiler plant in section 7.3.1. Note that directly fired units discharge the heat and products of combustion into the treated space.

— Identify any problem of stratification as this is common with convective heating systems in tall spaces. The temperature near a high roof or ceiling can be substantially higher than that at working level. Typical temperature gradients with different types of emitter are shown in Figure 10.5[4]. Low-speed fans or simple ductwork systems to recirculate warm air to low level can often be justified, unless there is a requirement for fume extraction at high level.

— Check damper positions and outlet terminals where air is ducted from units. Ensure that air is directed only to where heat is required and that discharge is not obstructed.

— Consider replacement with a radiant heating system if the air change rate in the treated space is high.

10.2.4 Electric space heating

In addition to purpose-designed systems, it is common to find small individual units in local use.

— Check existing controls for correct operation and settings.

— Check cleanliness of heat transfer surfaces and unit casings, freedom of air movement and freedom from obstruction.

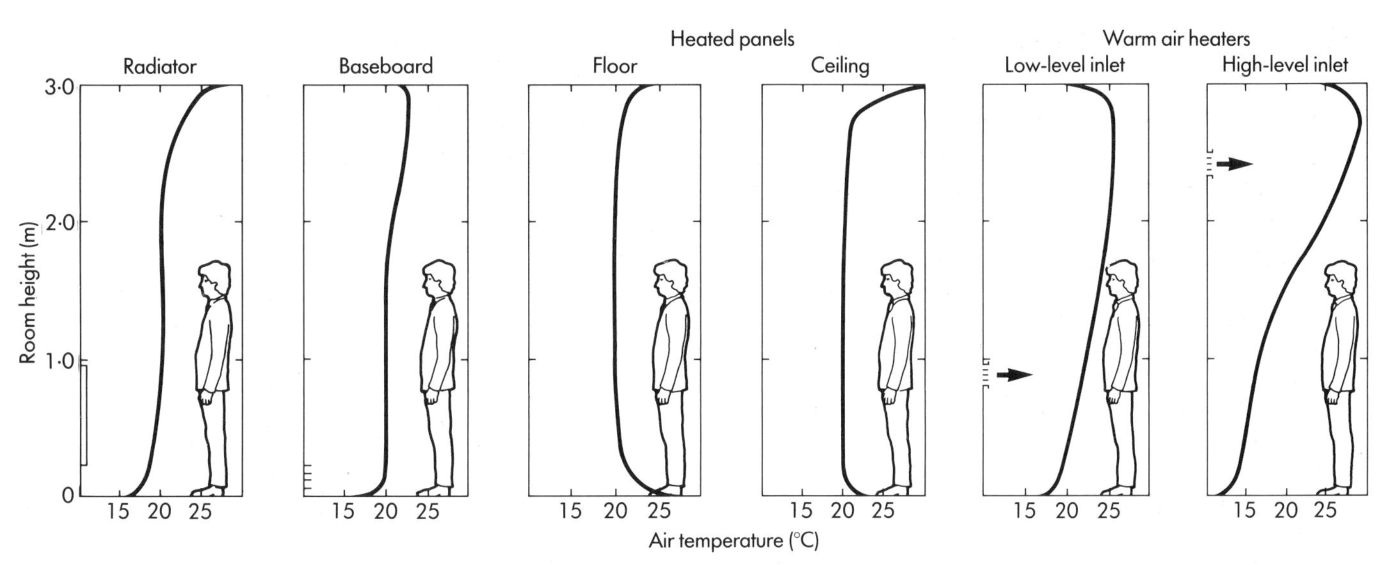

Figure 10.5 Vertical air temperature gradients[4]

- Consider installation of automatic controls, e.g. time-switches, room thermostats, occupancy detectors, for groups of heaters or individual units in regular use but switched manually at present.
- Pay particular attention to control of the charging of storage heaters. The charging should be reduced during mild weather and, if appropriate, at weekends. Automatic controls can often be applied to groups of heaters. Ensure that heaters are connected to a supply with an appropriate tariff.
- Consider alternative forms of heating that could offer lower operating costs than direct electric heaters. Storage heaters, gas-fired heaters and extension of an existing LPHW system are common alternatives. Note that local use of an electric heater may be cheaper than operation of a full LPHW system on some occasions, e.g. out of normal working hours.
- Ensure that effective manual control is maintained where this is most appropriate, e.g. for occasional, irregular use of individual heaters.

10.2.5 Temperature measurement for space heating systems

A digital thermometer equipped with air and surface contact probes and suitable for room and system temperatures is recommended for spot checks. Resolution to 0.1°C is useful when measuring space temperatures as the difference between inside and outside temperatures is relatively small. Consider the effect of a change in space temperature by just 1°C from 20°C to 19°C when the outside temperature is 8°C. Even this relatively small variation represents a change of over 9% in the temperature difference being calculated, i.e. 12 K compared with 11 K. As building heat loss is proportional to this temperature difference there is a corresponding effect on any calculation of energy requirement based on the measured values.

Fast response is also important for an instrument used for spot checks of different media, e.g. resolution to 0.1°C within 10 seconds when moving from ambient air temperature to hot water radiator temperature measurement. For air temperature measurements movement of the sensing probe at approximately 1 m/s can avoid radiant heat effects and improve the initial response time.

A comprehensive survey additionally requires the deployment of either thermographic chart recorders or solid-state data loggers. These are readily available for hire if outright purchase is not justified. A recording period of one week is generally found convenient and normally provides a satisfactory amount of information.

Chart recorders give an immediate visible record; data loggers can be downloaded to a computer for analysis. Whichever is used, the temperature profiles obtained can help to identify:

- when and where incorrect space temperatures occur
- whether system flow temperature is appropriately matched to outside temperature, when simultaneous recordings are made for comparison
- where equipment is operating for incorrect periods or is poorly controlled.

Figure 10.6 is an example record of space temperature in an office requiring a temperature of 19°C to be maintained between 0900 and 1700, five days per week. It can be seen that heating was not switched off at the weekend, that it was switched on too early and switched off later than necessary and that overheating occurred during occupancy.

Recording instruments must be located carefully if the results are to be interpreted correctly. Cold draughts, direct sunlight, proximity to heat sources and, by no means least, tampering can all produce misleading results. With caution the results of even a short-term recording can give a fair indication of what might be expected to occur over a full heating season.

10.3 Domestic hot water

10.3.1 Local water heaters

Direct supply units are usually located next to sinks and are either of the instantaneous supply or of the storage type. The energy source may be electricity or natural gas. For

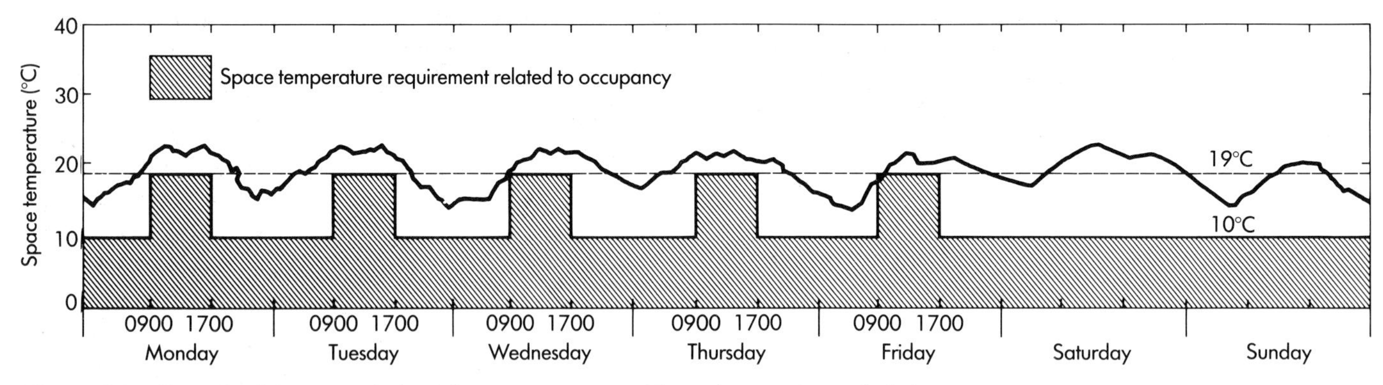

Figure 10.6 Example of thermograph chart of space temperature with requirement shown shaded

small individual units the potential energy savings are limited.

— Check for satisfactory water temperatures and flow rates (see section 10.3.2) and correct operation of controls.

— Examine the insulation of storage cylinders.

— Consider providing timeswitch control for larger storage cylinders, e.g. with immersion heater rating of 3 kW or more. The timeswitch program should aim to make maximum use of off-peak electricity and to minimise maximum demand charges.

— Check that heaters without timeswitch control are isolated manually when appropriate, e.g. over holidays or weekends.

10.3.2 Central systems

Domestic hot water may be generated centrally or in distributed heat exchangers supplied from a central boilerhouse with primary hot water or steam. There may be gravity fed or, in larger installations, pumped secondary circulation of hot water, or direct draw-off through dead legs.

— Review the need to operate large centralised systems, particularly during periods when only small localised users need hot water. The use of local fired water heaters or instantaneous units can sometimes produce major savings.

— Assess the distribution/standing losses. It may be possible to measure these by monitoring fuel consumption when the system temperatures are stable and the building is unoccupied, i.e. when there is no usage of hot water.

— Check the water temperature in storage and at the user end and ensure correct operation of controls. The hot water discharge temperature at the point of use should be 55–60°C for normal use; this could mean temperatures above 60°C at the point of storage. The water must be hot enough to minimise the risk of infection by *legionella pneumophila*, i.e. over 55°C, but not so hot as to result in excessive storage and distribution losses.

— Check the adequacy of storage cylinder and pipework insulation. Missing or damaged sections should be identified for early replacement or repair, but upgrading to increase thickness will probably not be justified. Recommended insulation thicknesses[5] range from 25 mm for the smallest pipe size to 75 mm for flat surfaces.

— Check that flow rates at outlets do not result in wastage. Consider the application of flow restriction devices to taps or replacement with percussion taps on wash handbasins. Touch- or occupant-sensitive timed flow controls might be recommended in multi-use locations where taps or showers are left running.

— Relate actual consumption to occupancy where make-up water is metered. The provision of metering should be considered for heavy users without this facility. Average daily consumptions from field trials[6] at various commercial premises are given in Table 10.9.

Table 10.9 Daily hot water consumption for various types of commercial buildings[7]; normalised assuming 65°C storage and 10°C cold feed temperatures

Building type	Total daily hot water consumption (litre/person)		Service (litre/person)	Catering	
				(litre/meal)	As % of total
Schools and colleges	Maximum	13	7	18	85
	Average	6	3	6	53
Hotels and hostels	Maximum	464	303	62	70
	Average	137	80	14	28
Restaurants	Maximum	17	10	73	95
	Average	7	3	8(4)†	60
Offices	Maximum	26	10	33	87
	Average	8	3	10	48
Large shops	Maximum	25	6	45	91
	Average	10	4	8	57

†4 litre/meal is the average consumption in restaurants without large bar facilities.

— Review storage capacities in relation to normal practice. The *CIBSE Guide*[7] suggests storage of approximately 1 litre per person for service and 2 litres per meal for catering in schools and offices with systems having a 1.5 hour recovery period.

— Consider replacing storage cylinders with plate heat exchangers to reduce standing losses.

— Identify dead legs with a view to eliminating them or reducing their length. Shortening runs reduces both draw-off and standing losses.

10.3.3 Temperature measurement of hot water systems

Spot checks of water temperature by digital thermometer are probably adequate for a concise survey.

The use of temperature recorders during a comprehensive survey of large distribution systems is recommended to:

— confirm how the system is operated, in terms of time and temperature control

— provide information on when significant draw-off occurs, reflected in the temperature profile

— indicate the recovery period on starting or after heavy draw-off.

10.4 Air conditioning and mechanical ventilation

Systems involving a sequence of processes must be examined closely to ensure that the best overall efficiency is achieved.

Figure 10.7 shows the basic processes which can occur as they might appear on a psychometric chart. In practice one or two will occur in sequence, e.g. cooling and dehumidifying followed by heating, to achieve the desired conditions.

The components of an air conditioning system can include:

— supply/extract fans

— heating/cooling coils

— humidifiers

— chiller plant

— condensers/cooling towers

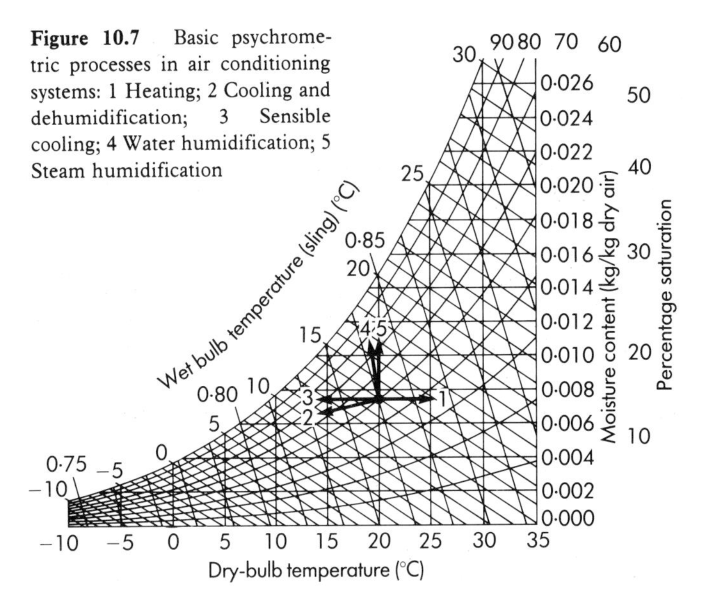

Figure 10.7 Basic psychrometric processes in air conditioning systems: 1 Heating; 2 Cooling and dehumidification; 3 Sensible cooling; 4 Water humidification; 5 Steam humidification

— control valves and dampers

— filters

— heat recovery devices, including heat wheels and run-around coils

— ductwork and terminals.

A survey should aim to identify the minimum acceptable extent to which air must be conditioned and the most efficient means by which this can be achieved:

— Establish the required conditions for the space(s) being served, including operational times, air change rates, dry-bulb temperature and relative humidity. It may be found that some conditioning processes are no longer necessary. Some typical design conditions are given in Tables 7.2–7.5.

— Examine the installed plant to establish the type of system, the forms of energy used, the processes performed and the condition of the components.

— Estimate the heating/cooling loads, taking account of incidental gains. Internal gains from equipment can be assessed from a knowledge of the ratings and diversity of use. In offices the average general power in use may be only a third of the nominal installed equipment load. Heat emissions[8] from occupants at various levels of activity are given in Table 10.10. Calculation of solar gain, involving details of location, orientation and glazing, is outside the scope of this Manual. Reference should be made to the *CIBSE Guide*[9].

Table 10.10 Heat emission from the human body (Adult male, body surface area 2 m^2[8])

Degree of activity	Typical application	Total $s + l$	Sensible (s) and latent (l) heat emissions (W) at the stated dry-bulb temperatures (°C)									
			15		20		22		24		26	
			s	l	s	l	s	l	s	l	s	l
Seated at rest	Theatre, hotel lounge	115	100	15	90	25	80	35	75	40	65	50
Light work	Office, restaurant†	140	110	30	100	40	90	50	80	60	70	70
Walking slowly	Store, bank	160	120	40	110	50	100	60	85	75	75	85
Light bench work	Factory	235	150	85	130	105	115	120	100	135	80	155
Medium work	Factory, dance hall	265	160	105	140	125	125	140	105	160	90	175
Heavy work	Factory	440	220	220	190	250	165	275	135	305	105	335

†For restaurants serving hot meals, add 10 W sensible and 10 W latent for food.

— In speculative developments in particular compare the actual loads with those specified by the developer. Systems designed and commissioned for inappropriate loads may need fundamental reappraisal. Studies[1] have found typical cooling load allowances for general power in offices to be between 5 W/m^2 and 13 W/m^2, depending on business type and building use.

— With major systems measure air volume flow rates, i.e. supply, extract, exhaust, recirculation and fresh air. Relate the measured values to building volume to determine air change rates and calculate the proportion of fresh air used. Compare with the actual requirements of occupancy. A minimum fresh air supply rate of 8 litres/s per person is recommended for sedentary occupants in a general building environment with no smoking. Statutory requirements and local byelaws normally apply to factories, theatres and toilet areas. Air flow volume rates should be quoted at reference conditions, e.g. 20°C and 1.013 bar.

— Measure building air conditions in terms of dry-bulb temperature and relative humidity and compare with requirements. Where anomalies are found, measure air conditions at terminal supplies and/or air extracts.

— For detailed surveys of major systems check air conditions at each stage of the air conditioning process, to ensure that excessive cooling or heating does not occur at any intermediate stage. Plotting the processes on a psychometric chart, as in Figure 10.7, will assist in understanding them.

— Ensure that the volume of air handled is the minimum necessary to meet all occupancy and statutory requirements while maintaining effective operation of the system. Savings can be made in fan power and in heating/cooling a reduced volume. This might be achieved by control of fan speed or blade characteristics. Restriction of air supply is a cheaper alternative but puts an added burden on the fan. Alternative volume flow rates may be appropriate for different occupancy patterns.

— Examine occupancy patterns and ensure that the system is switched off entirely whenever possible, e.g. overnight, at weekends or seasonally.

— Check the balance between supply and extract volumes. Pressurisation of a space results in a loss of treated air. A negative pressure can lead to infiltration of untreated air but may be required in some situations, e.g. laboratories. Ensure that maximum permissible use is made of recirculated air or free cooling with external air as appropriate. Overnight/early-morning precooling with untreated external air might be possible for further free cooling.

— Identify opportunities for recovering heat from exhaust air, particularly if recirculation is not possible. Ensure that heat recovery systems involving pumped circuits, heat wheels or other operating costs are only operated when economic.

— Review control arrangements and settings to ensure that controls allow maximum acceptable temperature with minimum cooling in summer and minimum acceptable temperature with minimum heating in winter. Check also that control interlocks are organised so that plant items do not run when there is no call for them, i.e. not always to fixed schedules.

— With dual-duct systems ensure that simultaneous heating and cooling are avoided by operating only the hot or cold deck at any given time. The other deck can continue to carry air which is neither heated nor cooled.

— Check that dehumidification is only operated when absolutely necessary in summer, since it involves cooling and subsequent reheating. This can be assisted by allowing relative humidity to rise to the maximum acceptable level; up to 70% should not cause discomfort in most environments.

— Ensure that heating load is kept to a minimum by avoiding humidification whenever possible in winter. Some organisations wishing to discontinue use of water sprays have found that room conditions remain acceptable without humidification and made savings in air heating and water treatment costs.

— Ensure that terminals, grilles, diffusers and induction nozzles are clean and free from obstruction.

— Consider opportunities to reduce the temperature of perimeter heating circuits used to meet fabric heat loss and prevent cold downdraughts, and increase the temperature of chilled water circulation to terminal or induction units required to meet the cooling load.

— Ensure that doors and windows of treated spaces are kept closed to avoid loss of conditioned air.

— Ensure that separate heating/cooling units and humidifying/ dehumidifying units or systems are not allowed to work in opposition in treated spaces.

10.5 Lighting

A review of the lighting requirements and the methods of control should be undertaken. Investigations should then consider how to make the maximum use of natural light. Artificial lighting should be made to harmonise with the available daylighting, using the most efficient appropriate source. The types of lighting system in use must be examined to identify where alternative, more efficient forms of lighting might be employed.

The safety and comfort of occupants must be taken into account at all times. Lighting has a direct effect on productivity and so it is important to ensure that satisfactory conditions are maintained.

(*a*) Establish the specific or recommended[11] requirements that determine the service to be provided, including any constraints on the types of fitting permitted:

- — general illuminance or task lighting
- — colour rendering; flicker-free source, avoidance of glare
- — flameproof/weatherproof luminaires
- — continuous or highly intermittent use
- — aesthetic qualities of lighting or equipment.

Requirements may have changed with the use of a building, e.g. greater use of personal computers and word processors requiring a lower level of background lighting. The standard service illuminances[11] for various activities are given in Table 10.11.

Table 10.11 Standard service illuminance for various activities/interiors[11]

Standard service illuminance (lx)	Characteristics of the activity/interior	Representative activities/interiors
50	Interiors visited rarely with visual tasks confined to movement and casual seeing without perception of detail.	Cable tunnels, indoor storage tanks, walkways.
100	Interiors visited occasionally with visual tasks confined to movement and casual seeing calling for only limited perception of detail.	Corridors, changing rooms, bulk stores.
150	Interiors visited occasionally with visual tasks requiring some perception of detail or involving some risk to people, plant or product.	Loading bays, medical stores, switchrooms.
200	Continuously occupied interiors, visual tasks not requiring any perception or detail.	Monitoring automatic processes in manufacture, casting concrete, turbine halls.
300	Continuously occupied interiors, visual tasks moderately easy, i.e. large details > 10 min arc and/or high contrast.	Packing goods, rough core making in foundries, rough sawing.
500	Visual tasks moderately difficult, i.e. details to be seen are of moderate size (5–10 min arc) and may be of low contrast. Also colour judgement may be required.	General offices, engine assembly, painting and spraying.
750	Visual tasks difficult, i.e. details to be seen are small (3–5 min arc) and of low contrast, also good colour judgements may be required.	Drawing offices, ceramic decoration, meat inspection.
1000	Visual tasks very difficult, i.e. details to be seen are very small (2–3 min arc) and can be of very low contrast. Also accurate colour judgements may be required.	Electronic component assembly, gauge and tool rooms, retouching paintwork.
1500	Visual tasks extremely difficult, i.e. details to be seen extremely small (1–2 min arc) and of low contrast. Visual aids may be of advantage.	Inspection of graphic reproduction, hand tailoring, fine die sinking.
2000	Visual tasks exceptionally difficult, i.e. details to be seen exceptionally difficult, i.e. details to be seen exceptionally small (< 1 min arc) with very low contrasts. Visual aids will be of advantage.	Assembly of minute mechanisms, finished fabric inspection.

(*b*) Identify the type and condition of the installation:

- Note the types and ratings of the lamps, control gear and systems.
- Note the cleanliness of lamps, diffusers, reflectors, photocells and other surfaces which affect the light output. Mirror reflectors fitted in multiple-tube fluorescent fittings may allow one or more tubes to be removed while retaining a satisfactory level of illumination. Without cleaning, the illumination available in a dirty environment could easily fall by half over just two or three years.
- Assess the general level of maintenance, including planned lamp replacement. The light output of discharge lamps, e.g. fluorescent tubes and SON lamps, falls with use — at least 20% deterioration is likely over three years. After 2000 hours use the output from a fluorescent tube is likely to fall by about 2–4% per 1000 hours.
- Note any opportunities for greater use of daylight. Ensure that windows, rooflights and room surfaces are kept clean. Sources of daylight should not be obscured unless there are overriding reasons to prevent glare or control solar gain and heat loss.

(*c*) Review the control arrangements:

- Lights should be switched off when not required; clearly labelled, accessible switches with reminder notices will encourage regular use. The addition of switches local to each user could also be considered as an aid to good housekeeping. Ideally, occupants should have to switch lights on manually but with some form of control for switching off.
- Group switching should be avoided where possible or arranged so that as few luminaires as possible are left on unnecessarily, e.g. adjacent to windows or in unoccupied areas. Switching of individual luminaires in intermittently used areas or of rows parallel to windows can be useful here.
- Selective switching for security or cleaning purposes can provide an alternative level of background lighting.

Table 10.12 Outline guidance on the application of lighting controls (source Property Services Agency)

Type of accommodation	Type of switching							
	User control			Automatic				
	Conventional wall switch	Individual pull cord switch on each luminaire	Hand held infrared or ultrasonic trigger switching individual luminaires	Timed bulk off switching	Photocell operated off switching	Daylight linked dimming	Acoustic and photocell switching	Range of potential energy savings (%) 'if correctly applied to suitable situations'
Open-plan offices	0	2	3	2	3	2	0	40–70
Cellular offices	0	1	3	2	3	0	R	40–70
Office with daylight	3	0	0	3	0	0	R	30–50†
Office without daylight	3	0	0	0(S)	0	0	0	10–20†
Large storage areas	0	0	3	3	2	1	R	40–60†
Factory areas (with daylight)	1	0	3	0	2	3	0(S)	40-60†
Sports halls	1	0	0	0	0	0	3	30–50†

Key
3 Preferred
2 Good
1 Poor
0 Inappropriate
0(S) Inappropriate on all luminaires for safety reasons
R Requires further investigation
† Estimated

— The following should be considered: occupancy detectors and/or delay timers for intermittently occupied areas; photocells where natural light is available and for outside lighting; timed controls to switch lighting off at the end of occupied periods; reset controls to allow individual lights to be switched back on locally. Guidance on the application of lighting controls[12] is given in Table 10.12.

(*d*) Measure the illuminance at workplaces or other points of interest and compare with the specified requirements. The standard instrument used in lighting measurement is an illuminance meter. Measurements should be aimed at identifying opportunities for reducing the lighting load. A considered approach is needed to obtain meaningful results:

— Lamps and meters should be allowed to stabilise before readings are taken. Note that colour correction factors have to be applied to the readings from some meters (in accordance with the manufacturer's instructions).

— Specific point measurements are necessary for fixed workplaces.

— An average of as many points as are needed to give a representative value should be taken to assess general background lighting.

— Daylight should normally be excluded during measurements of the illuminance provided by artificial lighting.

— Measurements taken with daylight available can suggest where alternative switching arrangements could allow greater advantage to be taken of natural light, e.g. switching off perimeter lights.

— Remember the possible effects of uncleaned surfaces and, with discharge lamps, total hours of lamp use. Even surrounding air temperature affects the light output from a fluorescent tube. The maximum output usually corresponding with a tube surface temperature of about 35°C. If illuminance is still adequate when output is restricted then there may be a case for reducing the installed load.

(*e*) Identify where and when the contribution from artificial lighting can satisfactorily be reduced:

— permanently, by removing or substituting alternative equipment

— occasionally, by selective switching to suit occupancy or available daylight

— by dimming

— by ensuring that lamps/luminaires/windows/building surfaces are kept clean and that discharge lamps are changed regularly.

(*f*) Explore the opportunities for changing the type of luminaire or lamp to reduce overall operating costs, particularly where systems are due for replacement or rewiring, or are used for extended periods. Allowance must be made for:

— Electrical costs, including demand charges. Note that a change in the type of load, e.g. conversion from tungsten to fluorescent lighting, can reduce the power factor (see section 8.1.6).

— Maintenance costs, including cleaning and lamp replacement. The reduced replacement frequency of long-life lamps is an important factor where access is difficult.

Typical schemes include:

— Replacement of tungsten filament lamps with fluorescent lamps; a 25 W rating compact fluorescent lamp gives a light output equal to that of a 100 W filament lamp and can sometimes be used directly in the same fitting. An example of the savings available by changing from tungsten to fluorescent lighting is given in Table 10.13.

— Replacement of standard-diameter fluorescent tubes with 25 mm diameter energy saving tubes; restricted to switch-start circuits and those with certain electronic starters. A change of tube type is recommended to be carried out when existing lamps fail or are scheduled for replacement.

- Replacement of tungsten filament or fluorescent lamps with fewer metal halide or sodium discharge lamps. This usually requires rewiring to new luminaire locations. It is not applicable to areas with low ceilings or where frequent switching is anticipated.
- Installation of local task lighting allowing use of fewer luminaires for general background lighting.
- Introduction of high-frequency operation, normally 28–30 kHz, for fluorescent lighting circuits; high-frequency control gear is most economical in twin tube or four-tube circuits. For a twin 1500 mm fitting with 38 mm tubes, the total circuit load can be reduced typically from 160 W to 111 W with a high-frequency system. High-frequency control gear also operates at a power factor near unity, unlike most conventional 50 Hz systems.

Schemes in which complete sets of new lamps, gear and reflectors can be fitted in existing housings offer the advantage of reduced installation costs with minimal damage and disruption.

Table 10.13 Example calculation of the savings available by replacing tungsten filament lamps with fluorescent lamps

Existing system	: 40 × 100 W tungsten lamps
Proposed system	: 10 × 1500 mm fluorescent lamps (26 mm krypton-filled)
Annual hours of use	: 5000
Average cost of electricity	: 6p/kWh
Annual energy cost existing system	: $40 \times \frac{100}{1000} \times 5000 \times 0.06$
	= £1200
Annual energy cost proposed system	: $10 \times \frac{70}{1000} \times 5000 \times 0.06$
	= £210
Potential annual energy saving (excluding savings in labour and materials due to reduced frequency of lamp replacement)	= £990

The total circuit load and initial light output of some standard types of lamp are given in Table 10.14.

10.6 Process and other equipment

10.6.1 Process plant

Process plant can be a major source of heat gains and is often implicated in cases of inefficient distribution systems. Consideration should be given to:

- **Distribution of services** — steam, hot water, insulation of hot pipework, incorrect sizing of pipes etc., leakages, bad drainage, use of incorrect pressures or temperatures, bad or non-existent metering, condensate return.
- **Cooling water** — extravagant use, possible recycling, use of cooling towers, re-use in other processes, incorrect filtration or chemical treatment.
- **Compressed air** — badly sized compressors, incorrect pressure, leakages, poor water removal, excessive oil contamination, waste in idling tools or machines, insufficient receiver volumes.
- **Plant fabric** — insufficiency of insulation generating excessive loads on building cooling and ventilation systems.
- **Waste heat recovery** — from cooling air or water, from drier exhausts, from vapours, from process liquids or solids, from hot gases leaving furnaces, etc. to provide domestic hot water or space heating. Destratification in tall spaces.
- **Electricity** — poor motor sizing; wasteful drives; inappropriate and wasteful lighting; local heating; low power factors; peaks in load patterns.
- **Ventilation** — excessive air changes in buildings due to poor design of fume and vapour extraction systems.
- **Waste disposal** — possible disposal of wastes and sludges by incineration and recovery of the waste heat to be recycled into factory heat requirements. Recovery of feedstocks from waste

10.6.2 Air compressors

Compressed air may be supplied for pneumatic control only, or may be used for handtools and other purposes.

Table 10.14 Typical load and output of various lamp types

Lamp type	Size	Circuit load (W)	Output (lumen)	Minimum expected useful life (hours)
Tungsten filament (GLS)	100 W	100	1200	1000
Fluorescent tube, 38 mm, white	1500 mm	80	5100	5000
Fluorescent tube, 26 mm (krypton)	1500 mm	70	5100	5000
Compact fluorescent	16 W	16–20	610–890	5000
Mercury vapour	80 W	93	3800	5000
Low-pressure sodium (SOX)	55 W	68	7300–7600	6000
High-pressure sodium (SON)	70 W	81	5500–5800	6000

Notes

Light output falls during life of lamp.

Efficiency (lumen output per watt) increases with rating for any given lamp type.

Useful life of filament lamps is limited by failure or, for discharge lamps, by reduction in output and economic factors; affected by frequency of switching, etc.

Exact figures for any particular product should be obtained from the relevant manufacturer.

Some of the following considerations also apply to other types of compressor, e.g. for refrigerant or process gases.

— Check that operating pressures and periods do not exceed system requirements. Some control systems may need a maintained air supply. Others may be operated intermittently or need only a storage reservoir to allow plant shutdown.

— Check loading of compressors; long periods on-load could indicate system leakage, long periods off-load are inefficient if compressor motors continue to run. In extreme cases a change of compressor type could be considered.

— Ensure that air compressors have a cool, clean supply of air, preferably from outside. A 4°C reduction in inlet air temperature will reduce the power requirement by approximately 1%. Precooling can sometimes be justified.

— Ensure that the air inlet is unobstructed and that filters are kept clean. A reduction in inlet pressure by 25 mbar due to restriction of the inlet, or prevailing atmospheric conditions, will increase the compressor power required by appropriately 2% for the same output.

— Check the compressor fittings and the distribution system, particularly flexible hoses and couplings, for leaks.

— Perform no-load tests by timing plant on load/off load to assess losses from distribution systems. With all users shut off, allow the compressor to run until full system pressure is reached and the compressor unloads. Time the pressure fall to the point where the compressor comes on load again. Then time the recovery of full pressure. Repeat this several times to give mean values. The air leakage Q_L can be calculated from the formula

$$Q_L = Q_D t_L (t_L + t_U)^{-1}$$

where Q_L is the leakage in litres of free air per second, Q_D is the compressor delivery capacity in litres of free air per second, t_L is the time loaded in minutes, t_U is the time unloaded in minutes.

The power required to deliver 10 litres of free air per second at 7 bar is in the region of 3 kW.

— Consider heat recovery either directly by ducting warm air from air-cooled plant or indirectly from the cooling water circuit where fitted. Aftercoolers, sometimes used to condense out moisture, and intercoolers on multi-stage machines can also provide opportunities for heat recovery.

— Consider applications for more efficient systems, e.g. hydraulics, as a replacement for compressed air.

10.6.3 Catering equipment

Catering is a substantial load in some establishments, often with equipment used intensively for regular, fixed periods. It can therefore have as significant an effect on maximum electrical demand as on energy consumption costs. There may be opportunities for replacing old equipment with more efficient modern equipment — microwave ovens, refrigerators, chillers. Heat recovery from extract hoods or dishwashers can also be considered, provided full attention is given to the likely contamination of heat transfer surfaces.

The main potential savings are likely to be identified in improvements in operating practice:

— Load equipment fully and avoid frequent part-load operation.

— Turn off equipment when not in use.

— Use lids, covers and doors on equipment.

— Keep refrigeration coils free of frost and ensure that condensers have free circulation of cool air.

— Locate chilled storage units clear of heating sources.

— Ensure adequate maintenance.

— Use hot feed rather than cold to dishwashers to reduce electric heating load.

— Use the most efficient appliance for each task.

— Discourage use of equipment to supplement space heating, by, for instance, ensuring that adequate background heating is provided at the start of the shift.

Metering of kitchen energy use for charging purposes is often recommended, particularly if facilities are hired out or operated by private contractors.

10.6.4 Motor-driven plant

Electric motors used for pumps, fans and other powered equipment can account for a substantial electrical load.

The efficiency of typical three-phase induction motors operating at full load ranges from about 75% for a 1 kW motor to 90% for a 30 kW motor. Single-phase motors are some 5% to 10% less efficient.

Efficiency is reduced at part load and below one-third load the drop is very significant. Many motors are run for substantial periods at part load; in fact some oversizing is inevitable with variable or cyclic loads. Motors rated above about 5 kW and in regular use may justify individual attention.

— Check the motor load, from measurements of voltage, current and power factor, and compare with the rating. The load drawn by a three-phase motor operating on a 415 V, 50 Hz supply can be calculated as follows:

$$\text{Load (kVA)} = \sqrt{3} \times \frac{415}{1000} \times \text{Running current (A)}$$

$$\text{Load (kW)} = \text{Load (kVA)} \times \text{Power factor (PF)}$$

The power input to a motor drawing a running current of 8 A and operating with a typical full load pf of 0.85 would be approximately 5 kW. Motors running at low load are likely to show power factors very much less than 0.85.

— Examine the actual load requirements and identify where load reduction might be possible, e.g. reduced pumping rates. Also check for correct alignment, belt tension and satisfactory lubrication.

— Identify opportunities for transferring motors to different loads to improve load matching.

— Consider the application of energy-saving, variable-speed controllers to motors operating predominantly at part load. Controls can increase part-load efficiency but may reduce full load efficiency. Speed control is also an effective method of varying pump and fan duties.

- Consider installing high-efficiency replacement motors. The extra cost in original equipment is recovered typically in 1–2 years.
- Identify where ancillary equipment is left running unnecessarily, including gland cooling circuits and DC generators for lift motors.

10.7 Building fabric

The design and construction of a building directly affect its thermal performance. When assessing performance, due consideration must be given to those factors which influence the energy requirements but about which little can be done:

- location: climatic conditions, exposure, influence of other buildings
- shape: high/low rise, deep/shallow plan, simple/complex
- size
- orientation
- age: moisture content of new masonry, degradation of old materials.

Measures can often be identified which can improve the performance of one or more elements of the building fabric. Prevention of excess air infiltration will not only reduce heat loss but is also likely to improve comfort. For a detailed assessment the principal routes of heat loss or gain must be modelled. Details of construction, dimensions and environmental conditions are required to calculate fabric and ventilation heat losses.

Heat loss through an element of the building fabric can be estimated approximately as:

$$Q_f = AU(t_i - t_o)$$

where Q_f is the energy loss through the fabric (W), A is the area of the element under consideration (m^2), U is the thermal transmittance (W/m^{2}K), t_i is the inside air temperature (°C) and t_o the outside air temperature (°C).

Ventilation heat loss from a given space can also be estimated approximately:

$$Q_v = \frac{1}{3} NV(t_i - t_o)$$

where Q_v is the ventilation heat loss (W), N is the number of air changes per hour, V is the volume of the space (m^3).

The annual fuel requirement is then:

$$F = Qh \frac{100}{\eta} \times 0.0036$$

where F is the fuel input (GJ), Q = $Q_f + Q_v$ (W), h is the annual number of hours of operation (h), η is the seasonal efficiency of the heating system (%).

Calculations based on steady-state characteristics cannot give an accurate picture of the real dynamic response of a building.

Opportunities for organising building usage more effectively might be identified where other measures are not possible. For example, relocation of staff or office equipment to an area with maximum heat loss and minimum solar gain could reduce the need for comfort cooling.

10.7.1 Roofs

- Check the thickness of insulation above ceilings in pitched roof spaces. Areas may be found where insulation is missing or defective. Old, compact insulation may have to be upgraded. Table 10.15 gives an example calculation of the savings achieved by adding roof insulation.
- Consider installing a suspended ceiling beneath flat or pitched roofs to reduce the transmittance and the treated volume. Fitting a ceiling to a partitioned area to enclose a small treated volume within a much larger untreated building volume could be most cost-effective.
- Consider direct application of insulation to the roof, particularly if the construction is single-skin sheeting. Upgrading the insulation of solid, flat roofs is rarely cost-effective, except during major refurbishment.

Note that additional ventilation, vapour barriers or other measures may be needed to prevent condensation in roof spaces. Insulation should not be allowed to obstruct ventilation, especially at the eaves of a roof. Services above an insulated ceiling may require protection against freezing. Electrical wiring in a confined, insulated space could need protection from overheating.

Values for average external temperature can be obtained from Figure 10.8. *U*-values for some common construction types are given in Appendix A9. The current *Building Regulations*[13] set standards for U-values of 0.45 W/m^2 for the walls and roofs of new buildings other than dwellings.

10.7.2 Walls

- Consider cavity wall insulation; pay particular attention to the requirements for suitable materials and professional installation procedures.
- Consider internal dry lining or direct insulation of the outer faces of external walls or those between heated/unheated/cooled areas. Beware of the risk of condensation in the outer skin of an external wall with internal insulation. Insulation of single-skin sheeting construction walls is likely to show the best justification. Otherwise, measures may be included with other refurbishment work.
- Identify problems of air infiltration through building openings, where services pass through walls or where dampers are left open unnecessarily. A significant proportion of infiltration can also be attributed to 'breathing' of the fabric in traditional UK structures.

Additional protection of services against condensation or frost damage may again be needed.

10.7.3 Windows

- Examine the condition of windows for maintenance requirements. Identify where work is needed to repair

Table 10.15 Example caculation of the energy saving available by installing 100 mm thickness loft insulation in a building with a pitched, tiled roof, loft space and 10 mm plasterboard ceiling

Roof area		520 m^2
Annual hours of operation		4500 h
Seasonal efficiency of heating plant		60%
Internal temperature maintained		19°C
Average external temperature		7°C
Initial U-value of roof		2.6W/m^2K
New U-value with 100 mm glass fibre quilt insulation		0.35W/m^2K
Existing annual heat loss	: $520 \times 4500 \times \frac{100}{60} \times (19 - 7) \times 2.6 \times \frac{0.0036}{1000}$	= 438 GJ
Annual heat loss with insulation	: $520 \times 4500 \times \frac{100}{60} \times (19 - 7) \times 0.35 \times \frac{0.0036}{1000}$	= 59 GJ
Annual evergy saving	= (438 − 59)	= 379 GJ

catches, replace broken panes, remove excess paint preventing closure, rehang or replace distorted frames.

— Consider double glazing or secondary glazing; these are most applicable if windows need replacement or if sound insulation is also required.

— Note the use of windows by occupants, who may need to be reminded of the good housekeeping value of keeping windows closed to minimise the space heating or air-conditioning loads as appropriate. Open windows during mild spells in the heating season can indicate that heating control is inadequate. In summer, there may be opportunities to avoid the need for mechanical ventilation or air conditioning by making maximum use of natural ventilation.

— Consider reducing glazed area, especially rooflights, provided any consequential additional cost of artificial lighting can be justified by the reduction in heat loss or improvement in comfort.

— Identify all opportunities for installing weather stripping. This can be an extremely cost effective measure in exposed locations.

— Encourage the use of curtains or blinds to reduce heat loss or, to a lesser extent, heat gain as appropriate.

— Consider applying window overhangs, other shading devices or solar film to control solar gain.

10.7.4 Doors

— Examine the condition of doors for maintenance requirements such as faulty catches or automatic closers, and misaligned or seriously distorted frames needing rehanging or replacement.

— Consider the application of automatic closers, air curtains, strip curtains, entrance lobbies, revolving doors, etc. to entrances that are often left open. Door closer mechanisms need to act rapidly but safely to minimise the time that a door remains open. Automatic controls which sense pedestrians or vehicles might be justified for large entrances.

— Consider providing personnel access doors to avoid frequent use of large vehicle access shutters or doors; these may offer safety benefits by separating pedestrians and vehicles.

10.7.5 Floors

— Note problems of air infiltration under or through suspended or ventilated floors; internal floor covering can be effective, provided the floor structure does not suffer adversely when ventilation is restricted.

— Consider covering solid floors to increase comfort.

— Consider insulation of floor slab if refurbishment is due.

10.8 Automatic controls

Many of the fundamental requirements of automatic control are common to a range of end uses. It is not the intention here to deal with any specialist theory in depth, rather to highlight the basic aspects which a survey ought to address:

— Is the user aware of what conditions are required?

— Does the user understand and make the best use of the available controls?

— Are the controls giving what the user needs?

— What instructions, alterations or additions to controls should be made?

The application of automatic controls is intended to assist in meeting building operational or environmental requirements in an effective and efficient manner. Where existing controls are adjusted or altered, the full implications of any change must be given proper consideration. New or replacement controls must be selected carefully to suit the particular application. Guidance relevant to specific end uses is given in the preceding sections.

Automatic controls should not be put forward as a solution to all the deficiencies of buildings or systems. Neither should they be seen as relieving staff of responsibility for controlling energy use. To be most effective, controls must be easy to understand and convenient to the user. Often a simple control system that is understood will be more effective than a complex system that is left to 'fend for itself'.

Reference to CIBSE Applications Manual *Automatic Controls*[15] is recommended.

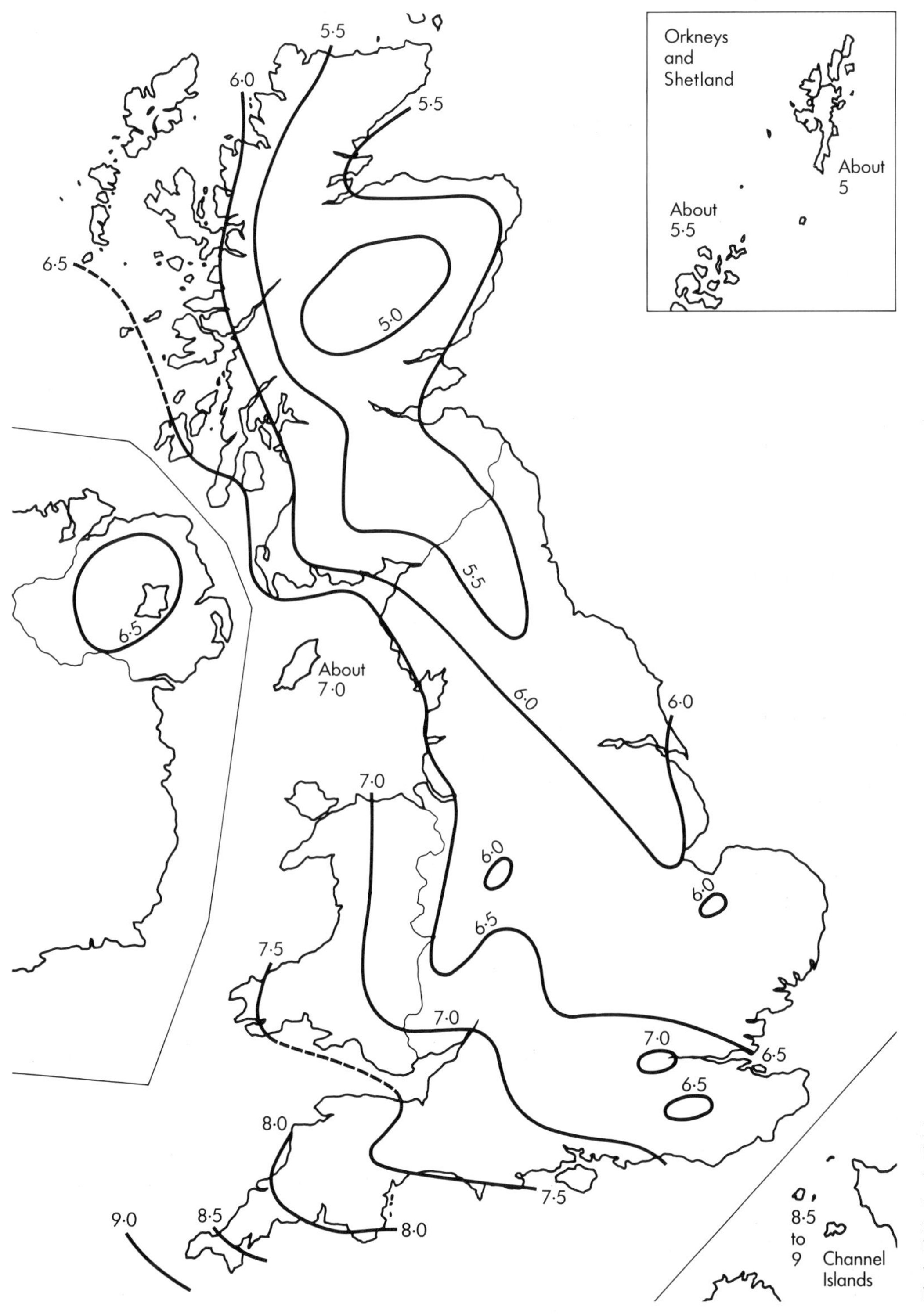

Figure 10.8 Seasonal mean dry-bulb temperatures[14]. For areas within a radius of one mile of the centre of large towns (population greater than 500 000) 0.5°C should be added, except for London, for which the value indicated should be used without modification.

10.8.1 Time control

Plant and equipment should be operated for the optimal periods to meet the requirements of occupancy and such other conditions as may be specified.

- Establish the precise pattern of occupancy in the areas served by a particular system; continuous, fixed daily or weekly pattern, highly variable, planned in advance, differences between areas on common system. Note any specific conditions to be maintained during or outside occupancy.
- Consider the type of equipment to be controlled. Main lighting systems should be in use only at times corresponding with occupancy. Heating systems can require variable preheat periods before occupancy, but can also be switched off before occupancy ends. Other plant may have to be available on demand. Boiler plant supplying both space heating and hot water services may have to be operated to meet both demands.
- Examine the controller in use for its suitability. A sufficient number and flexibility of ON/OFF events must be available for different start/stop times or day omission. Appropriate resolution and accuracy are needed. Three-hour time segments may be adequate or precise digital settings could be more appropriate. Switching times may have to be fixed or optimised. Lighting systems are often best switched off automatically at fixed times, but only switched on manually. Complexity should not be such that use is discouraged or that mistakes in settings are common. Extension timers can be provided for occasional occupancy outside normal hours. Night set-back of space heating can be practised where fully intermittent operation is not acceptable. Construction may need to withstand adverse conditions or prevent unauthorised tampering.
- Ensure that item controls are suitably located for the intended users. Local interval timers must be convenient for easy access; other controls may best be grouped in a central location to encourage full and proper use.
- Check the settings of the controller and observe the response of the system. Recordings of temperature over a day or a week can be used to identify where heating control settings should be changed, e.g. reduced preheat time. With optimum start controllers note that it is generally the occupancy times that must be set and not the plant switching times.
- Identify where additional time controls are needed to meet the requirements of different areas on a common system. Motorised zone valves can be considered for LTHW heating systems. Electrical circuits can be rewired for more flexible switching of selected luminaires or fan-assisted heating units. Alternative techniques, such as mains-borne signalling, can also allow selective switching.

10.8.2 Temperature control

Temperature controls must act in response to the demands of occupancy or other operational requirements, subject to the influences of uncontrollable factors such as ambient conditions and often within predetermined limits of comfort and safety.

- Establish the temperature requirements of each zone during and outside occupancy taking account of seasonal variations in comfort conditions (see Table 7.2). Review high/low limit and control settings. Note that set points are sometimes floating or reset in accordance with external influence. Actual needs may differ from stated requirements or original design intentions.
- Consider whether the parameters measured provide satisfactory temperature control. Dry bulb/wet bulb/ radiant temperature, solar/wind effects, water flow or storage temperature and duct air temperature are among the parameters that may need to be considered.
- Check that the location of sensors is representative of conditions. Space temperature sensors should be in a position not unduly influenced by direct heat gains or draughts but allowing free air circulation. 'Black- bulb' sensors for radiant heating should be located within line of sight of the heat source. Ensure that immersion, duct and surface sensors are fitted securely and identify where sensors and thermostats need additional protection to prevent tampering or accidental disturbance.
- Check that sensors have an appropriate range and resolution. Small errors can be significant, e.g. where the inside/outside temperature difference is small.
- Consider the suitability of the control type. Basic ON/ OFF control might be adequate. Some systems need a combination of proportional, integral and derivative control able to vary the rate of response to prevent overshoot and maintain stability.
- Check the settings and operation of the controls. Ensure that valves and dampers are free to operate. Compensating and reset controls may need to be checked over a range of ambient conditions. Recordings of temperature can assist in identifying where settings or controls need to be changed.
- Identify where additional controls are needed for central plant or complete zones. Heating circuit temperatures can be varied to suit weather conditions, unless they cannot be separated from a constant temperature requirement, e.g. fan convectors or hot water service. A compensator can be installed to control some gas-fired boilers directly in response to outside temperature if they have no minimum temperature restriction. Continuous automatic adjustment should be more effective than periodic manual adjustment of boiler thermostat settings. Pipework modifications may otherwise be necessary to maintain boiler flow temperature at say 83°C, while system temperature is varied down to perhaps 30°C by mixing/diverting arrangements.
- Identify the need for additional local controls. Thermostatic radiator valves could be installed for local trimming of space temperature in selected locations.

10.8.3 Frost/condensation protection

Intermittently operated heating systems should have an overriding control to provide protection from frost damage to pipework and condensation damage to the building fabric and contents during periods when space heating is normally off. The level of protection will be determined by, *inter alia*, the building type and contents.

— Check the location and setting of the frost/low limit thermostat. For an internal location, the *CIBSE Guide*[16] states that a low-limit value of 10°C is normal. An external thermostat with a setting near 0°C may be needed if there are external services to protect, but much higher settings would allow heating beyond that needed for purely internal protection. Replacement or relocation of existing thermostats should be considered if a lack of confidence in their accuracy or reliability has resulted in unnecessary heating. In a well insulated building the first stage of protection may only require pumps to be operated if circulation can prevent water circuits from freezing even without heat input.

10.8.4 Capacity control

Plant should be operated in such a way as to minimise standing losses associated with less efficient part-load operation or excessive cycling. Output has to be controlled to match the load, whether steady or variable. Central boiler, chiller and compressor plant require particular consideration.

— Compare the plant capacity with the anticipated load range. A single plant item which is grossly oversized may be suitable for derating to prevent excessive cycling. Multiple plant with different ratings might permit a more appropriate combination of plant to be operated. Smoothing of peak loads may avoid the need to bring extra plant on-line.

— Examine the methods for controlling the number of plant items on-line. Manual control tends to result in excess capacity being available. Reliance on automatic control can mean that opportunities for complete shutdown of standby plant in summer are missed. Full capacity may be required on starting.

— Consider the needs for standby plant. Complete isolation might be possible. Critical loads may demand instant availability. Facilities for restricting losses from standby plant could be appropriate, e.g. butterfly valves in boiler returns or flue gas dampers linked to burner operation.

10.8.5 Building energy management systems

Integration of the physical control of systems and plant with their overall management can allow more to be achieved than is possible with independent local controls and with less effort once a system is established. The potential for introducing, or extending, the application of intelligent automatic controls — building energy management systems — therefore deserves particular investigation. Opportunities for linking the processes of data collection, analysis, corrective action and reporting must be evaluated.

The benefits accruing to energy management can include:

— an interactive approach in which control settings can be changed and the results observed with minimum time and effort

— automatic remote data logging of consumptions, flow rates, environmental conditions, plant status, etc.

— integrated control of the systems on a site as a whole, particularly as applied to maximum demand control by load shedding or plant scheduling and optimisation of central plant related to distributed loads

— automatic analysis of logged data

— automatic calls for repair or condition maintenance

A more detailed feasibility study might be recommended to evaluate the application of BEMS where such benefits would clearly be advantageous, or when extensive adjustment/upgrading of existing controls is needed.

The extension of an existing system should also be considered. Those fortunate enough to have a system already can make use of the facilities for remote monitoring and data logging to assist directly in the survey.

10.9 Summary of section 10

* The understanding of energy use can be assisted by preparing a breakdown of consumption and energy flow diagrams as part of an audit.

* Data collected during a survey can assist in estimating the annual consumption of energy for various end uses for audit purposes.

* Space heating systems should be examined for their suitability, operation and control. Space temperature measurements require care in collection and interpretation.

* Domestic hot water systems can offer opportunities for savings in generation, storage and care. Correct temperatures must be achieved in water systems for health and safety.

* Air conditioning typically involves a series of processes that should be examined step-by-step.

* Lighting controls, lamp types and illumination levels can all be worth investigation.

* Process plant can cause significant local heat gains and have wider implications for site services.

* Building fabric has a direct influence on comfort as well as energy use. Air infiltration is often an important factor.

* Integration of the control of building services with energy management by means of building energy management systems may justify special study.

References for section 10

1 Owning and operating costs *CIBSE Guide* Section B18 (London: Chartered Institution of Building Services Engineers) (1986)

2 Calculation of energy demands and targets for new buildings *CIBSE Building Energy Code* Part 2(a) (London: Chartered Institution of Building Services Engineers) (1981)

3 Measurement of energy consumption *CIBSE Building Energy Code* Part 4 (London: Chartered Institution of Building Services Engineers) (1982)

4 Faber O and Kell J R *Heating and Air Conditioning of Buildings* rev. Kell J R and Martin P L (London: Architectural Press)

5 *BS 5422: 1977: Specification for the use of insulating materials* (London: British Standards Institution) (1977)

6 Jones P G The consumption of hot water in commercial buildings *Building Serv. Eng. Res. Technol.* **3** 95–109 (1982)

7 Water service systems *CIBSE Guide* Section B4 (London: Chartered Institution of Building Services Engineers) (1986)

8 Internal heat gains *CIBSE Guide* Section A7 (London: Chartered Institution of Building Services Engineers) (1986)

9 Summertime temperatures in buildings *CIBSE Guide* Section A8 (London: Chartered Institution of Building Services Engineers) (1986)

10 Leary J and Herridge S *General power in offices — the implications for HVAC design* (London: Electricity Council) (1987)

11 *CIBS Code for interior lighting* (London: Chartered Institution of Building Services Engineers) (1984)

12 *Energy management and good lighting practice* Fuel Efficiency Booklet **12** (London: Energy Efficiency Office) (1988)

13 *The Building Regulations 1985* (London: HMSO) (1985)

14 Weather and solar data *CIBSE Guide* Section A2 (London: Chartered Institution of Building Services Engineers) (1982)

15 *Automatic controls and their implications for system design* CIBSE Applications Manual **AM1** (London: Chartered Institution of Building Services Engineers) (1985)

16 Automatic controls *CIBSE Guide* Section B11 (London: Chartered Institution of Building Services Engineers) (1986)

11 Evaluating the survey findings

Survey findings must be evaluated and presented to whichever level of management has the authority to get the recommendations implemented.

11.1 Calculation of savings

The cost and energy savings anticipated from any measure can usually be based on:

- the estimated percentage reduction in annual consumption
- direct calculation from a reduction in the annual fuel cost, load, operating hours or energy loss.

For brief investigations aimed primarily at identifying opportunities, only initial budget estimates are expected.

Detailed surveys should, where possible, provide the data needed for more accurate calculations.

Savings in other areas, e.g. materials and labour costs, are not discussed here but should not be overlooked.

11.1.1 Assessed percentage savings

An arbitrary percentage saving may have to be assumed where an accurate figure cannot be calculated. For example, with good housekeeping or a formal energy monitoring and target setting system an energy saving of between 5% and 10% might be anticipated. Caution is needed in the application of such figures which should be based on experience and a sound understanding.

11.1.2 Direct calculation of energy savings

A saving should normally be calculated directly from a specific change in one or more parameters, e.g. different running time, change in temperature.

The calculation does not necessarily require a knowledge of the actual consumption of the item involved. For example, if a lighting load of 5 kW can be switched off for 2 hours every day then there will be a saving of 10 kWh per day. However, comparison with the total annual use will provide a useful check that the calculated saving is realistic.

11.1.3 Cost savings

Some cost savings, mainly relating to tariffs, may involve no actual energy saving.

The value of a saving in consumption is also dependent on the tariff applicable to the particular fuel involved:

Electricity

Each element of the tariff structure must be considered:

- The average overall unit price should only be used for approximate estimates concerning a reduction in unit consumption. If fixed standing charges are a significant factor or there is no corresponding reduction in any variable charges then such estimates could be misleading.
- Reduced unit consumption may give a saving in unit costs only if there is no anticipated effect on demand charges.
- Rescheduling of loads within a tariff band, without reducing unit consumption, might provide savings only in demand-related charges.
- The particular months or time of day involved often have to be taken into account.

Other fuels

Savings should generally be based on actual unit costs, excluding any standing charges, rather than overall costs.

11.2 Preparing the case for implementation

Technical and financial evaluation is needed to enable the options for savings to be put forward with confidence.

11.2.1 Technical considerations

The technical feasibility of recommendations should be confirmed:

- Ensure that implementation is physically possible, e.g. that space and access are adequate.
- Consider requirements for design, installation, supervision and future maintenance.
- Ensure compliance with health and safety regulations.
- Consider possible effects on other systems or operations.
- Check availability of proposed equipment/materials. Note whether there are restrictions on approved suppliers/materials.
- Indicate where there is any uncertainty associated with a particular aspect of any recommendation.

11.2.2 Financial viability

A financial appraisal of the economic benefits of each measure is needed to:

- evaluate the return available from the investment
- ensure that a positive return is available within the expected life of the measure
- compare the return with that of alternative measures or other opportunities for investment
- determine an order of priority for implementation

— plan the timing that will maximise the return on investment.

The method of calculation used to appraise the economic benefits of each measure must be appropriate to the investment involved. It should follow the procedures adopted by an organisation for any other investment. Opportunities will be missed if unfair criteria are set for energy saving measures.

Two principal methods of appraisal are considered here:

— Simple payback period

— Discounted cash flow (DCF).

The former is by far the commoner and more widely accepted method.

Simple payback period

Simple calculations of payback period are generally adopted for measures showing a return within five years or involving only very minor investment. The method is also appropriate for an initial assessment of measures that involve more substantial investment.

A calculation of simple payback period should be based on costs and savings as follows:

$$S = C/(F - O)$$

where S is the simple payback period (yr), C is the capital cost (£), F is the annual fuel cost saving (£), O is the additional annual operating cost (£).

It should be made clear whether costs and savings are based on firm quotations or budget estimates. Remember that design and implementation costs should also be taken into account.

If it appears unlikely that the criteria for investment can be satisfied then consider other benefits that may arise. Also consider whether a measure that cannot be justified in isolation could be worthwhile as part of a package of measures.

Discounted cash flow (DCF)

Large projects and long-term measures require the preparation of a cash flow statement to evaluate their true economic worth.

Discounted cash flow is a method of evaluating use to take account of the timing of capital and revenue costs and savings. Further details and examples of DCF calculation are given in Appendix A10.

The decision on when to apply DCF methods should reflect normal business policy. Further evaluation is considered to be beyond the scope of a survey and may have to take account of:

— tax allowances and liabilities

— the source and cost of funds

— risk and uncertainty.

11.2.3 Categories of recommendation

Recommendations, irrespective of their type, can usefully be grouped in categories according to the level of investment, namely:

— no/low cost measures

— medium cost measures

— high cost measures.

The definitions of low-, medium- and high-cost normally reflect the levels of expenditure for which different authority is required. For each cost category the recommendations can be summarised in a table of the form of Table 11.1. They should be arranged in a nominal order of priority for implementation.

Care is needed where implementation of one measure will reduce the savings available from subsequent measures. It must be made clear what allowance has been made for this effect as the implementation may actually take a course different from that anticipated. The order of implemen-

Table 11.1 Form used to summarise recommendations

Ref.	Location	Description of recommendation	Fuel type	Net annual savings		Implementation costs (£)		Simple payback (years)
				(£)	(GJ)	Extra study and design if applicable	Total	
Totals (and overall payback period)								

tation can be prioritised by cost and, if investment capital is limited, the ratio Net present value (NPV)/Investment cost is a good prioritising factor. Alternatively, Internal rate of return (IRR) prioritises according to the fastest return on investment. The terms NPV and IRR are explained in Appendix A10. Measures with a return higher than the basic profitability of the business should offer a sound financial case for implementation.

In addition to the positive recommendations arising from a survey a summary of the rejected options can be useful for future reference. Findings can be updated if a change in circumstances makes further review appropriate.

11.3 Presenting the findings

The importance of ensuring that the findings of a survey are understood and their value recognised cannot be overstressed.

Formal presentation is therefore recommended for all but the most basic of surveys.

A full report should include:

— observations supported by audit and test data

— conclusions and evaluation of the findings

— recommendations for action.

Suitable structures for survey reports are included in Appendices A4 and A5. The depth of technical discussion may vary widely in providing background to the recommendations, but there should always be a summary which can form the basis of a presentation to management.

11.4 Summary of section 11

* With careful technical and financial appraisal options for saving energy can be put forward with confidence.
* The criteria for investment in energy efficiency should be comparable to those applied to any other investment.
* Recommendations should be grouped in categories of capital cost and arranged in nominal order of priority for implementation.
* A formal presentation of survey findings and recommendations to senior management is recommended.

12 Implementation of survey findings

The principal object of the audit and survey stages is to identify worthwhile opportunities for saving energy. Production of a report must not be regarded as the end of a project but as a step leading to implementation.

Every effort must be made to ensure that a project proceeds to implementation. The complete process involves:

— final selection of the measures to be adopted

— planning and carrying out the selected measures

— monitoring the results.

12.1 Selection of measures

The final selection of measures is expected to be influenced strongly by the availability of funding and management time for implementation. Priorities are normally assigned on a financial basis, as discussed in section 11.2.

Preference might sometimes be given to establishing a programme of measures within the capacity of management to implement and support. A highly cost-effective measure may not be so attractive if savings are dependent on a continued management and maintenance input.

Options with a good potential for future development will also be favoured.

12.2 Implementing recommendations

A plan should be produced to implement the chosen options identifying:

— what action is necessary, e.g. design, tendering, installation, supervision, etc.

— who will carry out the tasks

— who should be informed or may be affected, including external bodies where relevant

— when, and with what priority, each measure will be implemented.

12.2.1 Collective plan

The plan developed will normally have to deal with measures in groups rather than individually.

For the purposes of implementation the best groups are likely to be based on the type of work involved, e.g. mechanical, electrical or insulation works. This should assist tasks such as tendering or installation, particularly if work is to be let to an outside contractor. Priorities can still be assigned to individual items within a given group by taking account of their costs and benefits.

12.2.2 Tasks involved

The tasks involved in implementing each measure could include:

— introducing changes in operating practice

— organising/providing training

— obtaining funding

— applying for planning permission

— preparing final designs or specifications

— inviting tenders or obtaining quotations

— reviewing financial viability after firm costs of measures are established

— making adjustments to existing equipment

— installing and commissioning new equipment

— supervising the works/project management

— setting up the mechanisms for managing/monitoring energy.

12.2.3 Allocation of tasks

Tasks should be allocated with the authority and support of senior management. It is recommended that someone within an organisation, normally the site energy manager, be given the role of project manager with responsibility for co-ordinating the works. Liaison with in-house staff at a range of levels and with external organisations may be required. Large organisations may wish to consider setting up an energy working party.

General good-housekeeping measures need the support and co-operation of all staff. In large organisations it may be necessary to resort to education or poster campaigns to ensure that everybody gets the message that they, too, are involved. Individuals within particular buildings or departments should be charged with local responsibility in this respect. Energy newsletters or articles in in-house publications can be effective in promoting energy awareness.

Proper co-ordination is particularly important where outside contractors, consultants or contract energy management organisations are used to assist in implementation. Information on the type of assistance available from consultants and contract energy management organisations is given in Appendices A2 and A3.

12.2.4 Consultation with interested parties

Consultation with safety officers, staff representatives, etc. is often necessary where working conditions or practices will be affected. Some measures may also have to be explained to building occupants if they are to be introduced successfully. Automatic lighting control is a good example of a measure that is most effective where it is understood and accepted by staff, but might otherwise fail. The additional effort involved in consultation is always worthwhile if it will improve results. Due time must also be allowed to meet any requirements for approval from outside bodies in connection with planning, insurance or health and safety.

12.2.5 Schedule for implementation

The short-term or low-cost measures can normally be given immediate priority for implementation. Investment measures often need to be assigned priorities which result in the best application of funds within available budgets. Apart from the availability of funding, the timing can also be influenced by the time involved in gaining approval, tendering or restrictions on when work can be carried out. Some measures, for instance, may not be possible until a planned seasonal shutdown.

The limiting factors should be identified and a programme drawn up to show when each measure is scheduled for implementation. Measures that have been deferred or retained only for future consideration should be included with a date for their review.

In a long-term programme, reinvestment of the revenue generated by savings from the first measures implemented can allow additional measures to be adopted, and is to be encouraged.

12.3 Monitoring of results

Good business practice requires that the results of any action be monitored. With monitoring, the effects of the measures implemented can be evaluated and, if appropriate, a case made for replication or further action to maximise the benefits.

Following implementation, monitoring should aim to identify:

— the energy savings made

— the true change in energy and other operating costs

— any other effects arising.

12.3.1 Energy savings

Overall consumption must be monitored in addition to that in any individual cost centres. This is necessary to confirm that savings in one area have not been made at the expense of increased consumption in another. It is not unknown for supplementary electric heating to be found where improved control and insulation have reduced beneficial heat gains.

— Confirm which metered and monitored areas of consumption on which the implemented measures should have an effect.

— Following implementation, identify any trends in consumption which can be attributed to the measures undertaken.

— Evaluate the energy and cost savings actually achieved and compare with the predicted values.

— Investigate the reasons for any significant variance from expectations — this could be due to incorrect estimates, to factors not allowed for at the estimating stage or to deficiencies in the method of monitoring.

Organisations operating energy monitoring and target setting (M&T) systems should be in a position to review the immediate progress made when measures are implemented. It will be necessary to update the standards and targets against which energy consumption is regularly measured to take account of the predicted savings, or to set completely new targets.

12.3.2 Other effects

In addition to the main quantitative analysis of energy use the monitoring process should allow for a qualitative assessment of the effects of implementing recommendations on operations. This may in turn become quantitative if productivity can be measured.

— Where a change in environmental conditions is anticipated from adjustment or replacement of controls or plant, make spot checks or recordings of the new conditions achieved. Ensure that the measures have been adopted satisfactorily and that controls and equipment are not being misused or overridden. Review the effects during periods of occupancy and vacancy.

— Measures to improve fabric insulation or reduce infiltration heat loss can result in high space temperatures. Ensure that controls are adjusted correctly either to prevent this effect or to maintain it within required limits.

— Take note of the reaction of the users of buildings or plant and determine whether activity/productivity has changed. If no reaction occurs this may indicate successful introduction of a measure, but there could still be scope for further improvement.

— In the case of process plant ensure that there has been no unacceptable effect on product quality of product or the ability of plant to meet production demands.

— Review the above items periodically. Carry out a formal assessment each year to assist interpretation of the results of the annual audit.

— Finally, remember that the whole survey process should be repeated typically every three to five years to maintain progress and ensure that new opportunities are not missed.

12.4 Summary of section 12

* An action plan should be produced, identifying the tasks involved in implementing each measure, the allocation of these tasks and the programme for implementation.

* Consultation with staff representatives, building occupants and other interested parties can assist with the introduction of measures and may be vital to their success.

* The results of implementing measures should be monitored regularly, in line with good business practice.

Appendix A1: Glossary

Audit

A review of energy use and costs normally performed in conjunction with a site investigation.

Availability charge

The charge made for maintaining an agreed electrical supply capacity to a consumer's premises.

Blowdown

The removal of accumulated solids in water at the base of a steam boiler by drawing off a small proportion of that water.

Building energy management system (BEMS)

A computer-based system of remote control and monitoring of building services used for interactive energy management.

Coefficient of performance (COP)

For a refrigeration system, the ratio of useful cooling effect to the total power input to the system.

Combined heat and power

Combined generation of electricity and heat.

Consultant, energy

An individual or organisation able to offer independent and professional advice or services to an organisation wishing to implement an energy audit and survey.

Contract energy management (CEM)

A service providing technical, financial and management resources to implement an energy saving project. Remuneration for the service is often by retention of a proportion of the savings. The CEM contractor can also bear a higher proportion of the financial risk of any investment.

Cost centre

An area or department for which energy costs can be determined and related to the activity or other operating costs of that area.

Discounted cash flow, net present value (NPV)

The value of cash inflows less the value of cash outflows over the life of a project, with all future cashflows discounted back to present day values. A positive result is needed for a project to be acceptable.

Discounted cash flow, internal rate of return (IRR)

The discounting rate which gives a 'breakeven' result i.e. zero net present value. It can be compared with other investments or the cost of borrowing.

Energy-efficient building

One which provides the specified internal environment for minimum energy cost, normally within the constraint of what is achievable cost effectively.

Energy manager

A person in an organisation with responsibility for energy matters.

Energy user

Any organisation using energy.

Gas supply, interruptible

A supply of gas which may be interrupted within specified limits at the discretion of the supply company, allowing lower contract prices than a firm supply for large users.

Gas supply, firm

A supply of gas available continuously.

In-house expertise

Knowledge and experience held by persons within an organisation.

Maximum demand (MD)

Maximum power, measured in kW or kVA, supplied to a customer by a supply/distribution company, equal to twice the largest number of kWh or kVAh consumed during any half-hour in a specified period (usually a month). Charges for maximum demand usually vary seasonally.

Measures, good housekeeping

Actions that can be taken to save energy requiring no capital expenditure.

Measures, low-cost

Energy saving measures requiring minimal capital expenditure.

Measures, medium-cost

Investment measures that involve a medium level of capital expenditure.

Measures, high-cost

Measures involving major capital expenditure that may need further study and authorisation at executive level.

Monitoring and target setting (M&T)

A method of energy management in which real energy consumptions are recorded regularly and related to specific variables to allow comparison with the standard or target values calculated. Corrective action is then taken where appropriate.

Performance indicator (PI)

The value of annual energy consumption related to a building or site characteristic; most commonly gigajoules per square metre of floor area under consideration.

Power factor

The ratio of kW/kVA in AC electrical circuits relating useful power to reactive power. Values significantly below unity may attract penalties from the supply authority.

Regional Energy Efficiency Officers

Specialist members of staff within the regional offices of the Department of Trade and Industry, the Welsh and Scottish Offices, and the Northern Ireland Department of Economic Development; available to advise organisations on all aspects of energy efficiency.

Sankey diagram

A diagrammatic representation of the flow and use of energy through a site, building or system.

Simple payback period

The time calculated to recover an investment when the capital cost of implementing a measure is divided by the net annual saving.

Survey, comprehensive

A detailed site investigation of specified aspects of energy use providing firm recommendations for energy- saving measures.

Survey, concise

A short site examination of specified aspects of energy use to identify potential energy saving measures.

Appendix A2: Consultancy

A2.1 Choice of consultant

A clear idea is needed of the tasks to be performed so that the services available from a consultant can be matched to the project requirements. The particular combination of services required and the scale of a project determine the type of consultancy likely to be found of best value.

Small energy consultancy practices often provide only an energy audit and survey service. The client then seeks separate services from suppliers and contractors to design and implement the schemes recommended in the survey report. Other consultancies may concentrate on design services, or might offer a specialist survey and design service for a particular type of plant or building. Larger consultancies can be expected to offer the broadest range of services and to be able to maintain support throughout all stages of a project from initial planning to final implementation.

The greatest benefits are obtained when a consultant is chosen and briefed carefully. When choosing a suitable consultant, consideration of the following points will enable a commission to be placed with confidence:

- **background of consultancy**; it is best to use a well established business that has tackled similar projects and can demonstrate long-term commitment.
- **references** from recent clients or direct previous experience of the services offered.
- **details** of the experience, qualifications and supervision of individuals who actually carry out the commission.
- **resources**; ability to meet the programme; ability to meet temporary instrumentation requirements.
- **independence** from interests in supply of equipment or fuel that could compromise impartiality of advice.
- **fee bases**; fixed sum, time charge, scale fee, percentage or performance related charges may be preferred.

Advice on where energy consultancies can be contacted may be obtained from professional institutions and Regional Energy Efficiency Officers. Further information is available from the Energy Systems Trade Association (which has established the Independent Energy Consultants Group).

A2.2 Use of model brief

If the intended scope of a commission has to be determined then preliminary discussions should be held and, perhaps, outline proposals invited from prospective consultants. Clients should satisfy themselves that their requirements are covered and that no unnecessary items are included.

Formal proposals should detail the fees involved, the objectives, procedures and reporting format, and indicate the anticipated benefits. To ensure that the terms of reference are clearly defined, and to assist with the comparison of proposals from alternative consultants, the use of a standard form of brief is recommended.

The models for comprehensive and concise audit and survey briefs given in Appendices 4 and 5 respectively are intended for this purpose. Deletions or additions of specific items should be made to match the precise needs of a particular site.

A2.3 Use of consultants

To make the most effective use of a consultant's time it is necessary to identify the information and assistance that the consultant can expect to receive, and to ensure that maximum co-operation is available. If the consultant knows with confidence what support is being provided then work can be priced accordingly and effort directed properly.

The findings and recommendations of a consultant are rightly challenged so as to ensure confidence in the results of an audit and survey and establish a firm case for implementation. Findings are most likely to be implemented successfully where the parties have worked together and maintained a constructive dialogue.

Appendix A3: Contract energy management

A3.1 Types of service

Contract energy management (CEM) companies can provide differing combinations of the financial, engineering and management services needed to implement an energy saving project including:

- a site survey to establish the requirements and produce proposals for energy-saving measures including control adjustments, energy management practices, replacing or upgrading equipment and new techniques. The contract costs and potential energy savings will also be assessed.
- management of project implementation from design to commissioning.
- financing of capital works.
- service and maintenance of equipment and supervision of operation for period of contract.
- fuel purchasing for the contract period.
- performance monitoring for the contract period.

A3.2 Basis of contract

The contractual arrangements used by CEM companies normally fall into one of three categories. Heat service contracts involve variable charges for energy supply. Shared savings contracts involve variable charges based on a share of the savings achieved by implementing energy efficiency measures. Fixed-fee contracts involve fixed charges for a given level of service.

With a heat service contract the CEM company effectively acts as a utility company. There is no incentive for the CEM company to reduce the user's energy demand. These contracts are most appropriate to process energy users or where plant condition and operation are problematic.

Shared-savings contracts are likely to result in the greatest savings. The CEM company will identify energy saving measures and often provide full assistance with implementation and funding. Improvements involving both the supply and use of energy might be realised. Costs and savings require calculation procedures but the user may be offered some guarantee of savings and the risk is likely to remain low. The procedures are most readily applied to heating and air conditioning services that can be modelled to assess the savings.

For a fixed fee, a CEM company may offer to take over the operation and maintenance of building services and pay for the energy used. The user will gain advance knowledge of the service costs but have no incentive to reduce energy use. Such contracts are most common in the commercial sector where suitable manpower might be limited.

Each contract type should provide a guarantee that under no circumstances will the total contract charges exceed the costs that would arise if there were no CEM agreement.

A3.3 Application of CEM

CEM may be appropriate where in-house resources are inadequate to implement a project and it is wished to minimise the financial risk associated with the project.

Suitable sites range from single small office blocks to large industrial complexes. The contract and types of measure implemented can be matched to individual site requirements. Contracts might involve major plant replacement but schemes involving control changes, thermal insulation, lighting replacement, energy management advice, monitoring and target setting and on-going review of suitable technologies are often included.

It is not uncommon for energy management systems to be installed, where justified by the benefits and the scale of the contract, allowing the CEM company to monitor performance and results from their own premises.

Organisations lacking only the finance for capital investment or, alternatively, only the availability of technical expertise may prefer to find other means of providing those resources. Those requiring a more comprehensive package from a single source may wish to consider CEM.

A3.4 Stages in adopting CEM

With contract sums expected to be related to historic energy costs it is in the client's interest to ensure that savings available from simple no-cost or low-cost measures are already being achieved. The audit and survey needed to identify the basic measures should also reveal the potential for higher-cost measures and, if carried out independently, will provide the client with additional information of value when considering proposals for CEM.

The particular skills, experience and contract types offered by different CEM companies should be considered in relation to the plant and equipment involved. Clients should also satisfy themselves that the levels of maintenance and service offered match their requirements. As long-term contracts are normally involved it will be necessary to confirm the stability of the CEM company and the provision for responding to changes in circumstances. The precise mechanisms by which different contracts operate must then be reviewed on their individual merits.

CEM companies will usually perform a preliminary survey at their own expense before submitting proposals or entering contract negotiations. An estimate of overall savings should be available at this stage.

Typically, a cost will then be agreed with the selected company to undertake a full study and prepare a detailed report. The cost may be absorbed into the contract or covered by a once-off payment if the contract is not continued further. Direct use of a model survey brief will not normally be appropriate.

Further guidance is available in the CIBSE Applications Manual AM6: *Contract Energy Management* (1991).

Appendix A4: Model brief for a comprehensive energy audit and survey

This Appendix outlines the objectives of a comprehensive energy audit and survey, and the report format to be used to detail the findings and recommendations of the audit and surveys.

A4.1 Objectives

The objectives of a comprehensive energy audit and survey are to

- provide an audit of site energy
- identify areas of potential energy cost savings
- provide an estimate of potential annual energy savings with implementation costs and payback periods
- identify how methods of energy management should be developed to achieve, maintain and recognise further potential savings.

Methods of achieving these objectives are by:

- analysis of invoiced and metered fuel consumptions
- observations and measurements on energy-consuming equipment during the survey period to determine energy efficiency and wastage
- examination of operating practices and management techniques
- establishment of a basis for continued monitoring of energy consumptions and setting achievable targets
- on completion of the survey, preparation of a report in the format outlined below containing recommendations supported by data, which, if implemented, would result in energy costs savings
- presentation of the report to senior management.

A4.2 Report format

A report is written to detail the findings and recommendations arising from the survey. It consists of the following.

A4.2.1 Management summary

This outlines the potential energy savings identified by the survey. The summary shall also show tables of individual recommendations based on good housekeeping and low-cost measures and capital expenditure schemes.

A4.2.2 Report

The report details findings and recommendations in the following sections: (Delete any sections to be excluded.)

- Site information
- Energy audit
- Energy use:
  - (*a*) Boiler plant
  - (*b*) Space heating
  - (*c*) Domestic hot water
  - (*d*) Air conditioning and ventilation
  - (*e*) Electrical power and lighting
  - (*f*) Catering
  - (*g*) Other energy uses
  - (*h*) Building fabric
- Energy management.

A4.2.3 Appendix

The Appendix includes graphs, calculations and miscellaneous data which are relevant to the report.

A4.3 Scope of survey

The following items should be covered and reported on (Delete any items to be excluded).

A4.3.1 Site information

The site, its functions and services, shall be described in this section. A building or site plan shall be included.

A4.3.2 Energy audit

Based on information obtained from fuel invoices, metered consumptions, observations and calculations, the following shall be produced:

- a table showing the consumptions, unit costs, and total costs for all purchased fuels for the previous 12 months.
- a table showing the percentage changes in energy costs over the previous 3 years.
- a summary of energy intakes e.g. supply meters and tariffs.
- table(s) and pie chart(s) showing a breakdown of fuel types and costs for each major fuel user, for the previous 12 months.

Energy performance indicators for the building(s) shall be calculated and commented on.

A4.3.3 Energy use

Boiler plant

- Combustion efficiency tests shall be carried out on all boilers at high, medium and low fire rates (where applicable). Recommendations for the improvement of combustion efficiency shall be made where necessary.

— Seasonal efficiency of boilers shall be estimated, based on observed operating conditions and past records. The effect of producing hot water only, during summer periods, on seasonal efficiency shall be assessed if applicable.
— Where low seasonal efficiencies are found, recommendations shall be made on savings achievable by replacement boilers, separate hot water heaters, or other facilities for seasonal operation.
— The general condition of the boilers shall be assessed with particular reference to insulation and air inleakage, and recommendations made for improvements where necessary.
— The condition and thickness of insulation on pipework, valves and flanges shall be assessed and recommendations made for improvement where necessary.
— Suitability and settings of time and temperature or pressure controls shall be assessed and recommendations for improvement made where necessary.
— The use of waste heat recovery from boiler blowdown on steam boilers and the use of economisers on gas-fired boilers shall be evaluated and recommendations made on the viability and practicality of such schemes.
— The use of cheaper or alternative fuels shall be considered.
— Consideration shall be given, in certain circumstances, to the viability of the use of waste incinerators.

Space heating

The heating systems shall be examined and recommendations made on:

— the heating periods compared with occupancy
— the type of heating system installed
— the condition, settings, positioning and operation of existing controllers and sensors
— the need for additional controls
— the condition and thickness of insulation on pipework, valves and flanges
— the condition, positioning, and any obstruction of heat emitters
— a temperature record, over a period of 7 days, shall be carried out to obtain a heating profile in representative area(s).

Domestic hot water

The hot water system shall be examined and recommendations made on:

— calorifier storage and delivery temperatures and control systems
— calorifier insulation, condition and thickness
— calorifier storage capacity in relation to draw-off requirements
— hot water outlet temperatures and flow control from taps and showers

Electrical power and lighting

Observations and measurements shall be carried out to determine:

— the 'maximum demand' profile over a period of 7 days
— the most economical supply tariff, based on the 'maximum demand' profile and past invoices
— the need for power factor correction, if low power factors are penalised by the supply authorities
— excessive transformer losses due to low loading of transformers
— an assessment of connected power and lighting loads
— the type, condition, siting and switching arrangement of existing luminaires, and possible replacement by high-efficiency lamps
— any unnecessary use of lighting and power equipment, with particular attention to electric heating equipment
— the control of electric heating, and possible replacement by other types of heating
— the performance and loading of air compressors, and the potential for waste heat recovery
— the effectiveness of the distribution and utilisation of compressed air
— the type, size and loading of motors to suit the application
— the potential for combined heat and electricity generation (CHP).

Recommendations shall be made on the basis of the findings from the above observations and measurements.

Air conditioning and ventilation

— Air flows shall be measured and the heating/cooling loads assessed.
— Controls shall be examined for switching, settings and operation of existing time and capacity controls.
— Ductwork shall be examined for leakage and for correct operation of dampers.
— The performance and loading of refrigeration compressors and the potential for waste heat recovery shall be studied.
— The distribution and insulation standards of the refrigeration system shall be examined.
— The power consumption of major fans, chillers and pumps shall be assessed for audit purposes.

Catering and other energy use

— The use of cheaper alternative fuels and heat recovery shall be considered.
— Any unnecessary use of equipment shall be identified.

Building fabric

The following shall be examined and appropriate recommendations made:

- insulation standards of roofs, walls and floors
- glazing standards of windows
- excessive air infiltration due to badly fitting doors and windows
- excessive air infiltration due to doors and windows being left open

A4.3.4 Energy management

An assessment shall be made of the existing energy management procedures, information available, and metering at the site. Recommendations shall be made on any improvement which can be made to the existing system. These recommendations shall take account of manpower availability and the cost requirements setting up an improved system of energy management.

Systems of energy management shall be based upon quantitive measures of performance using:

(*a*) Performance indicators calculated for the particular situation

or

(*b*) Monitoring and target setting, relating energy consumption to known variables.

The merits of suitable methods shall be outlined on the basis of the availability of information, cost of implementation, metering availability and the suitability of variables against which to compare energy consumption.

Management structures for collecting and processing data and taking action in response to the findings shall be reviewed.

Appendix A5: Model brief for a concise energy audit and survey

This Appendix outlines the objectives of a concise energy audit and survey, and the report format to be used to detail the findings and recommendations of the audit and survey.

A5.1 Objectives

The objectives of a concise energy audit and survey are:

- to identify opportunities for reducing energy costs
- to estimate the potential savings, and where applicable, implementation costs
- to provide an audit for the site on the basis of the previous 12 months' invoiced accounts.

Methods of achieving these objectives are:

- by observations and, where applicable, analysis of how efficiently energy-consuming equipment is being used
- by considering possible improvements to energy management control.

A5.2 Report format

A short report shall be written to outline the findings and recommendations arising from the survey. The report shall be preceded by a summary outlining the potential energy savings available at the site. These will primarily be of the good housekeeping and low-cost type but will also indicate where further opportunities may exist. The body of the report shall contain the following sections:

- Site information
- Energy audit
- Energy use
- Energy management.

A5.3 Scope of survey and report

The following shall be covered:

A5.3.1 Site information

The site, its functions and services, shall be described.

A5.3.2 Energy audit

Based on data obtained from the previous 12 months' fuel invoices, a table showing annual fuel consumptions and costs shall be compiled for the site. Performance indicators shall be determined and commented on.

A5.3.3 Energy use

Boiler plant

Combustion efficiency, based on waste gas analyses, shall be assessed for the main boiler plant under operating conditions as found. The general condition of the boiler plant and associated pipework insulation shall be checked. Recommendations for improved energy efficiencies within the boiler house shall be based on the above analysis and observations.

Space heating and domestic hot water

The heating and hot water systems shall be assessed and recommendations made on:

- the heating period compared with occupancy periods
- the condition, settings and siting of existing controllers and sensors
- instantaneous temperature measurement taken during occupancy periods
- the condition of insulation on pipework, valves and flanges
- the condition and siting of heat emitters and any obstruction
- HWS temperature.

Electrical power and lighting

Observations of power and lighting systems shall be carried out to determine the following:

- the condition of lighting equipment
- any unnecessary use of lighting
- the type of existing luminaires and possible replacement by higher-efficiency lighting
- use of electric heating and its control
- the operation and loading of refrigerators and air compressors
- efficient use of large electric motors.

Recommendations to reduce energy costs shall be made on the basis of the above observations.

Air conditioning plant

The settings of existing time and capacity controls shall be obtained and included in the report, together with comments on control, operation and potential energy savings.

Building fabric

Observations shall be made of:

- insulation standards
- excessive air inleakage into buildings due to badly fitting doors and windows.

Recommendations shall be based on the above observations.

A5.3.4 Energy management

Existing energy management procedures shall be assessed, and outline recommendations shall be made on any improvement which can be made to the existing system. An assessment of any potential for fuel and/or tariff changes shall be outlined.

Appendix A6: Energy performance yardsticks

The following data are taken from the Energy Efficiency Office's *Energy Efficiency in Buildings* booklets to which reference is recommended. The use of performance yardsticks is summarised in section 5.2.2.

Table A6.1 Performance yardsticks for schools (kWh/m^2 per year)

Type of school	Energy efficiency rating		
	Good	Fair	Poor
Nursery	<370	370–430	>430
Primary, no indoor pool	<180	180–240	>240
Primary with indoor pool	<230	230–310	>310
Secondary, no indoor pool	<190	190–240	>240
Secondary with indoor pool	<250	250–310	>310
Secondary with sports centre	<250	250–280	>280
Special, non-residential	<250	250–340	>340
Special, residential	<380	380–500	>500

These values include schools with and without their own kitchen for providing meals on the premises. Schools with central catering facilities serving several other schools in the area are not covered, and this factor should be allowed for separately.

Table A6.2 Performance yardsticks for catering establishments (kWh/m^2 per year)

Type of establishment	Energy efficiency rating		
	Good	Fair	Poor
Restaurants	<410	410–520	>520
Public houses	<340	340–470	>470
Fast-food outlets	<1450	1450–1750	>1750
Motorways service areas	<880	880–1200	>1200

Alternative yardsticks

The fast-food outlet yardstick was calculated from a sample of premises with an average floor area per seat of 2.1 m^2. Where premises have a significantly higher or lower ratio it is recommended that the alternative yardsticks be adopted.

Since the level of trade will have a significant effect on energy consumption, a performance indicator based on floor area alone may not fully reflect the energy efficiency of catering establishments. If your energy efficiency is indicated as being poor then the alternative yardsticks given in Tables A6.3 and A6.4 will indicate whether this is due to higher than average trade. The energy consumption and number of meals or bar sales must correspond to the same period of time (preferably the same 12-month period).

Table A6.3 Alternative performance yardsticks for restaurants and fast-food outlets (kWh/meal)

Type of establishment	Energy efficiency rating		
	Good	Fair	Poor
Restaurant (a la carte menu)	<9.4	9.4–10.5	>10.5
Fast food outlets	<1.4	1.4–1.9	> 1.9

Table A6.4 Alternative performance yardsticks for public houses (kWh/composite barrel of beer)†

	Energy efficiency rating		
	Good	Fair	Poor
Public houses	<190	190–265	>265

†The number of composite barrels of beer is determined from bar sales as follows:

Number of composite barrels = No. of gallons of beer/36 + No. of gallons of wines and spirits/3.

Table A6.5 Performance yardsticks for shop sales floor area (kWh/m^2 per year)

Type of shop	Energy efficiency rating		
	Good	Fair	Poor
Department/chain store (mechanically ventilated)	<520	520–620	>620
Other non-food shops	<280	280–320	>320
Superstore/hypermarket (mechanically ventilated)	<720	720–830	>830
Supermarket, no bakery (mechanically ventilated)	<1070	1070–1270	>1270
Supermarket with own bakery (mechanically ventilated)	<1130	1130–1350	>1350
Small food shop — general	<510	510–580	>580
— fruit & veg.	<400	400–450	>450

Note

The above yardsticks relate to heated and naturally ventilated shops, except where indicated differently.

Table A6.6 Average percentage of total floor area given over to sales for the samples of Table A6.5

Type of shop	%
Department/chain store	51
Other non-food	67
Superstore/hypermarket	61
Supermarket no bakery	43
Supermarket with bakery	52
Small food — general	70
— fruit & veg.	69

Table A6.7 Alternative performance yardsticks for shop total floor area (kWh/m^2 per year) (See note on p 82.)

Type of shop	Energy efficiency rating		
	Good	Fair	Poor
Department/chain store (mechanically ventilated)	<250	250–280	>280
Other non-food shops	<170	170–200	>200
Superstore/hypermarket (mechanically ventilated)	<440	440–500	>500
Supermarket, no bakery (mechanically ventilated)	<470	470–560	>560
Supermarket with own bakery (mechanically ventilated)	<560	560–620	>620
Small food shop — general	<360	360–400	>400
— fruit & veg.	<280	280–310	>310

Note to Table A6.7

If non-sales areas are substantially larger than the average of our sample and are heated/air conditioned and illuminated to the same level as sales areas, then your performance indicator could be higher than the yardsticks given in Table A6.5. If this is the case, an assessment of your performance indicator may be made on a 'total area' basis. These yardsticks may also have to be adjusted if your shop has air conditioning or mechanical ventilation.

Table A6.8 Performance indicators for further and higher education establishments (kWh/m² per year)

Type of establishment	Energy efficiency rating		
	Good	Fair	Poor
Universities	<325	325–355	>355
Polytechnics			
Residential buildings	<230	230–315	>315
Teaching and administrative buildings	<190	190–260	>260
Colleges of Further Education	<230	230–280	>280

Notes

1 The university yardsticks combine residential and academic buildings as these are generally not separately metered (campus site with central boilerhouse). The average proportion for residential buildings is 25%. Most universities have some air conditioning (up to 10% of floor area in the sample) and mechanical ventilation (typically 50% of floor area). The yardsticks include these energy usages.

2 Separate yardsticks are available for residential and non-residential buildings at polytechnics as these tend to be separately metered buildings (i.e. no central campus). University residential accommodation should have a similar energy yardstick to that for residential buildings at polytechnics.

Table A6.9 Performance yardsticks for office buildings (kWh/m² per year)

Type of site	Energy efficiency rating		
	Good	Fair	Poor
Air-conditioned offices			
over 2000 m²	<250	250–410	>410
2000 m² and less	<220	220–310	>310
Computer centres	<340	340–480	>480
Naturally ventilated offices			
over 2000 m²	<230	230–290	>290
2000 m² and less	<200	200–250	>250

Note

The yardsticks for air-conditioned offices apply where at least 60% of the total floor area is air-conditioned. For offices where less than 60% is air-conditioned and the remainder heated and naturally ventilated, an appropriate yardstick Y_{pac} should be determined as follows: $Y_{pac} = (a/60)\,Y_{ac} + (1 - a/60)\,Y_{nv}$ where a is the percentage of floor area served by air-conditioning, Y_{ac} is the fully air-conditioned yardstick, Y_{nv} is the naturally ventilated yardstick.

Table A6.10 Performance yardsticks for sports facilities (kWh/m² total floor area per year)

Type of sports facility	Energy efficiency rating		
	Good	Fair	Poor
Swimming pool	<1050	1050–1390	>1390
Sports centre with swimming pool(s) (Pool/Total area less than 20%)	<570	570–840	>840
Sports centre without a swimming pool, or sports club	<200	200–340	>340

Table A6.11 Alternative yardstick for swimming pools (kWh/m² pool total surface area per year)

	Energy efficiency rating		
	Good	Fair	Poor
Swimming pool	<4900	4900–5900	>5900

Table A6.12 Performance yardsticks for public buildings (kWh/m² per year)

Type of building	Energy efficiency rating		
	Good	Fair	Poor
Library	<200	200–280	>280
Museum or art gallery	<220	220–310	>310
Church	<90	90–170	>170

Table A6.13 Performance yardsticks for hotels (kWh/m² floor area per year)

Type of site	Energy efficiency rating		
	Good	Fair	Poor
Small hotels and guesthouses (51 m²)†	<240	240–330	>330
Medium-sized hotels (61 m²)†	<310	310–420	>420
Large hotels (58 m²)†	<290	290–420	>420

†Average gross floor area per hotel bedroom.

Notes

1 These yardsticks are for hotels which are open throughout the year.

2 A performance yardstick for a hotel with an indoor heated swimming pool may be derived by combining the separate yardsticks for hotels and swimming pools in proportion to their respective areas.

Table A6.14 Alternative performance yardsticks for hotels (kWh/bedroom per year)

Type of site	Energy efficiency rating		
	Good	Fair	Poor
Small hotels and guesthouses	<14 000	14 000–19 000	>19 000
Medium-sized hotels	<19 000	19 000–25 000	>25 000
Large hotels	<15 000	15 000–21 500	>21 500

Table A6.15 Performance yardsticks for high street agencies (kWh/m² per year)

Type of site	Energy efficiency rating		
	Good	Fair	Poor
Banks and Post Offices	<180	180–240	>240
Building societies	<130	130–170	>170
Estate agents	<175	175–260	>260
Insurance brokers	<140	140–210	>210
Travel agents	<165	165–245	>245

Note

The samples of building societies, insurance agents, estate and travel agents premises used to produce the above yardsticks were predominantly all-electric and included only limited amounts of air-conditioning and comfort cooling.

Table A6.16 Performance yardsticks for entertainment buildings based on building volume (kWh/m^3 per year)

Type of site	Energy efficiency rating		
	Good	Fair	Poor
Cinemas (11.0 m)†	<59	59–71	>71
Theatres (15.0 m)†	<40	40–60	>60
Bingo clubs (11.0 m)†	<57	57–70	>70
Social clubs (3.8 m)†	<53	53–95	>95

†Average floor-to-ceiling heights. If the building height differs significantly from these values, use a more appropriate figure to convert from kWh/m^3 to kWh/m^2 or vice versa.

Table A6.17 Alternative yardsticks for entertainment buildings based on floor area (kWh/m^2 per year)

Type of site	Energy efficiency rating		
	Good	Fair	Poor
Cinemas	<650	650–780	>780
Theatres	<600	600–900	>900
Bingo clubs	<630	630–770	>770
Social clubs	<200	200–360	>360

Table A6.18 Performance yardsticks for public sector buildings (kWh/m^2 per year)

Type of site	Energy efficiency rating		
	Good	Fair	Poor
Prisons	<550	550–690	>690
Police stations	<340	340–470	>470
Fire stations	<440	440–620	>620
Ambulance stations	<400	400–530	>530
Crown and County Courts	<220	220–300	>300
Transport depots	<310	310–380	>380

Table A6.19 Performance yardsticks for factory and warehouse buildings (building volume basis) (kWh/m^2 per year)

Type of site	Energy efficiency rating		
	Good	Fair	Poor
Factories, with little or no process energy requirement			
– small, less than 2000 m^2†	<29	29–38	>38
– large, more than 2000 m^2	<33	33–46	>46
Factories, with heat gains from manufacturing plant‡			
– small, less than 2000 m^2	<24	24–34	>34
– large, more than 2000 m^2	<26	26–38	>38
Warehouses, heated	<19	19–34	>34
Cold stores¶	<55	55–75	>75
Hangars	<15	15–55	>55

†Small low-energy factory units with a high standard of insulation should only use about 85 kWh/m^2 for space heating.
‡The yardsticks exclude energy used for the process energy itself.
¶The cold store yardsticks exclude energy use for freezing down produce on site. As an approximate guide, add 250 kWh for each tonne frozen per year.

Note

The average heights assumed for the above building types are: Factories 8.0 m, Warehouses 8.0 m, Cold stores 9.0 m, Hangars 14.5 m.

Table A6.20 Alternative performance yardsticks for factory and warehouse buildings (floor area basis) (kWh/m^2 per year)

Type of site	Energy efficiency rating		
	Good	Fair	Poor
Factories, with little or no process energy requirement			
– small, less than 2000 m^2†	<230	230–300	>300
– large, more than 2000 m^2	<260	260–370	>370
Factories, with heat gains from manufacturing plant‡			
– small, less than 2000 m^2	<190	190–270	>270
– large, more than 2000 m^2	<210	210–300	>300
Warehouses, heated	<150	150–270	>270
Cold stores¶	<500	500–675	>675
Hangars	<220	220–800	>800

Appendix A7: Degree–days

Degree–day data give a measure of the average outside temperature, which has a direct influence on heating or cooling loads. The data are normally used for analysis of monthly energy consumption.

Monthly and annual figures are published in many sources[1,2] and should be adequate for an audit. Table A7.1 gives average figures for the last 20 years in seventeen regions for comparison with current data, which should be obtained month by month. An average of the two nearest regions, or data recorded on site, may be preferred for locations remote from the central recording location in any one region.

The figures in Table A7.1 are heating degree days calculated to a base outside temperature of 15.5°C, the base suitable for most commercial and industrial buildings. Figures to an alternative base of 18.5°C, commonly referred to as hospital degree–days, are published and used by the Department of Health. The higher base temperature is more suited to hospitals and other buildings with low internal heat gains and extended periods of heating to at least 19°C.

Table A7.1 20-year average degree-day data to base 15.5°C

Month	Region																
	Thames Valley	South Eastern	Southern	South Western	Severn Valley	Midland	West Pennines	North Western	Borders	North Eastern	East Pennines	East Anglia	West Scotland	East Scotland	North East Scotland	Wales	Northern Ireland
Jan.	346	368	345	293	321	376	361	375	376	381	372	378	383	388	401	330	365
Feb.	322	344	327	285	305	359	340	345	349	358	352	349	352	357	368	320	334
March	286	312	301	271	280	322	312	323	330	322	313	317	328	332	346	307	320
April	205	233	229	207	201	243	230	245	271	247	232	239	246	263	277	240	242
May	120	150	148	137	128	162	144	167	206	168	154	149	170	197	206	170	171
June	51	74	72	63	56	83	75	90	117	87	78	73	94	109	120	92	92
July	22	39	39	28	24	44	38	50	66	46	42	40	58	62	74	49	53
Aug.	25	44	43	28	27	48	39	56	68	49	44	39	64	67	78	45	59
Sept.	54	82	79	55	61	90	78	96	104	88	81	71	111	109	127	77	99
Oct.	130	160	150	116	138	178	157	171	182	175	165	154	188	192	203	145	173
Nov.	242	267	251	206	237	275	267	284	282	281	272	269	299	301	311	235	282
Dec.	312	334	312	258	300	343	328	341	339	346	341	341	352	354	362	294	329
Total	2115	2407	2296	1947	2078	2523	2369	2543	2690	2548	2446	2419	2645	2731	2873	2304	2519

References for Appendix A7

1 *Building Serv.: CIBSE J.* (London: Building Services Publications) (Published monthly)

2 *ENTEL Energy Information Service* (Prestel 50033)

Appendix A8: Correction of meter readings

A8.1 Natural gas

Care should be taken to ensure that data are related to standard, purchased conditions. Volume varies by approximately 1% for a gas temperature change of 3°C and by approximately 1% for a pressure change of 10 mbar (4″ water gauge). Minor variations from the standard conditions of 15.5°C and 1013.25 mbar can usually be ignored, but where gas is metered at high pressure/temperature the temperature correction factor f_t and pressure correction factor f_p are given as follows:

$$f_t = \frac{273 + 15.5°C}{273 + t_a}$$

where t_a is the actual temperature (°C).

$$f_p = \frac{P_a + 1013.25}{1013.25}$$

where P_a is the actual pressure gauge reading (mbar).

As an example, consider a meter supplied with gas at 16°C and a gauge pressure of 600 mbar.

To correct the meter data to standard conditions the overall correction factor will be:

$$\text{CF} = \frac{273 + 15.5}{273 + 16} + \frac{600 + 1013.25}{1013.25}$$

$$= 1.59$$

A8.2 Fuel oils

Fuel oils are sold by volume and not on an energy basis. The calorific value of oils may usually be obtained from the supplier. The heat value may be in kJ/l or kJ/kg. In the latter case, conversion to heat per unit volume requires a knowledge of the oil density in kg/l, and this value may also be obtained from the supplier. Again, the density is quoted at the standard temperature, 15.5°C.

The oil temperature at the metering points may well be as high as 120°C for the heaviest grade. The volume changes by approximately 1% when the temperature changes by 15°C. This effect is slightly more significant for lighter grades of oil, but these are not heated to the same extent.

For residual fuel oils, the approximate temperature correction factor by which meter readings should be multiplied to give standard volumes is:

$$\frac{1}{1 + 0.0007T}$$

where T is the actual oil temperature (°C) minus 15.5°C.

For example, the volume of a heavy fuel oil metered at 110°C should be corrected by the factor:

$$\frac{1}{1 + 0.0007\,(110\text{–}15.5)} = 0.938$$

Furthermore, if the above oil had a density of 0.98 kg/l and a quoted calorific value of 41 130 kJ/kg the calorific value per standard litre would be simply 41 130 × 0.98 = 40 307 kJ/l.

Where the consumption of the heavier fuel oils is determined by the delivery and tank stock method, minor errors will occur if a float-type contents gauge is used due to volume changes with temperature. However, tank gauges which measure the weight head of oil are calibrated for each grade of oil density and are therefore self-compensating.

A8.3 Coal

Coal sub-meters, for example on individual boilers, normally operate volumetrically and therefore require calibration for the particular size of coal in supply. The time taken to consume a delivery of known mass can provide a measure of the firing rate during a test.

A8.4 Liquefied petroleum gas

LPG is usually sub-metered volumetrically and therefore corrections similar to those for natural gas must be made.

A8.5 Steam

It is essential that the meter be calibrated for steam at the average supply pressure (and hence density). If steam is supplied at a lower pressure than the calibration pressure the meter will read ‘fast’, and conversely at higher pressure it will read ‘slow’. Steam density data may be obtained from standard tables. The correction factor CF for changes in steam pressure is given as follows:

$$\text{CF} = (\rho_a/\rho_c)^{1/2}$$

where ρ_a is the actual and ρ_c the calibration density (both in kg/m³).

Appendix A9: *U*-values

Table A9.1 Typical *U*-values for basic construction types (normal exposure) [1]

Building element	*U*-value (W/m^2K)
Windows	
Single glazing, 10% area wood frame	5.3
Single glazing, 10% area aluminium frame (no thermal break)	6.0
Double glazing, 10% area wood frame	3.0
Double glazing, 10% area aluminium frame (no thermal break)	3.6
External walls	
220 mm brickwork, unplastered	2.3
220 mm brickwork, 20 mm glass fibre quilt, 10 mm plastered	1.0
100 mm lightweight concrete block, 25 mm air gap, 105 mm brickwork, 13 mm lightweight plaster	0.88
100 mm lightweight concrete block, 50 mm UF foam, 105 mm brickwork, 13 mm lightweight plaster	0.45
Asbestos-cement cladding sheets	5.3
Asbestos-cement cladding sheets, 75 mm rigid insulating board	0.4
Roofs	
5 mm asbestos cement sheet, pitched	6.5
10 mm tile, loft space, 10 mm plasterboard ceiling	2.6
10 mm tile, loft space, 100 mm glass fibre quilt, 10 mm plasterboard ceiling	0.35
19 mm asphalt, 150 mm aerated concrete slab, 13 mm dense plaster	1.9

1 Thermal properties of building structures *CIBSE Guide* Section A3 (London: Chartered Institution of Building Services Engineers) (1980)

Appendix A10: Discounted cash flow

Discounting is a useful process for comparing costs and benefits which occur at different times. The benefits of investment measures are expected to arise in later years than their initial costs. An appraisal of alternative options may have to compare expenditure now with different savings achieved in future.

Most organisations place greater weight on earlier rather than later costs and benefits. A discount rate is therefore needed to determine the present-day value of cash flow in future years.

To separate the discounting procedure from inflation the discount rate is set in real terms. The real or test discount rate (TDR) is then the required rate of return over and above inflation, i.e. equal to the annual rate of return less the inflation rate. Each organisation should require a rate of return on energy efficiency measures consistent with its overall investment programme.

The chosen test discount rate is applied to the cash flow in each year of a project in the form of discount factors (see Table A10.2).

As an example, with a TDR of say, 7%, a £1 saving in one year's time, has a present value of only:

$$£1 \times \frac{100}{107} = £0.935$$

The discount factor in this case is 0.935 and would be applied to all cash flow in year 1 of projects with a TDR of 7%.

The sum of the discounted savings or income from a project less the discounted expenditure is called the net present value (NPV).

Consider a project involving an energy saving measure that has a capital cost of £3000 and that will produce an estimated annual saving of £800. The full capital expenditure occurs at the start of the project but the savings will provide an income in each year of the life of the project, say five years.

A10.1 Net present value

Assume that a test discount rate of 7% is required. The incomes for each year must be multiplied by the relevant discount factors from Table A10.2 and summated to give the cumulative discounted annual income. Capital expenditure is then deducted from this total to give the net present value (Table A10.1).

Table A10.1 Example of net present value calculation

Year no.	Income (£) × discount factor	Discounted income (£)
0	(Capital expenditure)	(3000.00)
1	800 × 0.935	748.00
2	800 × 0.873	698.40
3	800 × 0.816	652.80
4	800 × 0.763	610.40
5	800 × 0.713	570.40
	Net present value	280.00

Provided that the NPV remains positive the project is considered worthwhile. In the example the calculated value of £280 is positive and the case for investment is clearly favourable.

When the price of energy is increasing at a higher rate than inflation then the future income will be even greater and the case for investment all the better.

Table A10.2 Discounting factors

Future years	Discounting rate (%)										
	5	6	7	8	9	10	11	12	13	14	15
1	0.952	0.943	0.935	0.926	0.917	0.909	0.901	0.893	0.885	0.877	0.870
2	0.907	0.890	0.873	0.857	0.842	0.826	0.812	0.797	0.783	0.769	0.756
3	0.864	0.840	0.816	0.794	0.772	0.751	0.731	0.712	0.693	0.675	0.658
4	0.823	0.792	0.763	0.735	0.708	0.683	0.659	0.636	0.613	0.592	0.572
5	0.784	0.747	0.713	0.681	0.650	0.621	0.593	0.567	0.543	0.519	0.497
6	0.746	0.705	0.666	0.630	0.596	0.564	0.535	0.507	0.480	0.456	0.432
7	0.711	0.665	0.623	0.583	0.547	0.513	0.482	0.452	0.425	0.400	0.376
8	0.677	0.627	0.582	0.540	0.502	0.467	0.434	0.404	0.376	0.351	0.327
9	0.645	0.592	0.544	0.500	0.460	0.424	0.391	0.361	0.333	0.308	0.284
10	0.614	0.558	0.508	0.463	0.422	0.386	0.352	0.322	0.295	0.270	0.247
11	0.585	0.527	0.475	0.429	0.388	0.350	0.317	0.287	0.261	0.237	0.215
12	0.557	0.497	0.444	0.397	0.356	0.319	0.286	0.257	0.231	0.208	0.187
13	0.530	0.469	0.415	0.368	0.326	0.290	0.258	0.229	0.204	0.182	0.163
14	0.505	0.442	0.388	0.340	0.299	0.263	0.232	0.205	0.181	0.160	0.141
15	0.481	0.417	0.362	0.315	0.275	0.239	0.209	0.183	0.160	0.140	0.123

A10.2 Internal rate of return (IRR)

The IRR is effectively the TDR which gives a cumulative discounted annual income over the project life equal to the capital expenditure. A discount factor therefore has to be found which will make the NPV zero. For the example project this is found from Table A10.2, by a trial, to be between 10% and 11%, as shown in Table A10.3.

Table A10.3 Example of net present value calculations used to determine internal rate of return

TDR (%)	Year no.	Income (£) × discount factor	Discounted income (£)
	0	(Capital expenditure)	(3000.00)
	1	800 × 0.909	727.20
10	2	800 × 0.826	660.80
	3	800 × 0.751	600.80
	4	800 × 0.683	546.40
	5	800 × 0.621	496.80
		NPV	32.00
	0	(Capital expenditure)	(3000.00)
	1	800 × 0.901	720.80
11	2	800 × 0.812	649.60
	3	800 × 0.731	584.80
	4	800 × 0.659	527.20
	5	800 × 0.593	474.40
		NPV	43.20

If the figure of 10% over and above inflation exceeds that required for any investment then the case for the project is favourable.

Different investment opportunities can also be compared fairly using this method.

Appendix A11: Water surveys

Water is often grouped with energy as a utility and is commonly included as part of an energy audit or survey. It should not be regarded as an unavoidable overhead as there is much that the consumer can do to ensure value for money in water and sewage charges. Potential cost savings can be comparable to those available in energy cost savings.

For most consumers water cost savings of the order of 10% are often easily identified. Substantially higher percentage savings are not uncommon for small consumers.

A water survey should follow a series of procedures similar to those applied to energy:

— Collect consumption and cost data. These may be based on historical rateable value or metered values. Distribution drawings will also be useful, particularly if a site has multiple metering.

— Identify major trends which need to be explained or investigated.

— Conduct an audit to identify end users. If the likely consumption can be estimated then a case for changing to a metered tariff or a smaller meter size might arise. Comparison of predicted and actual metered use can also indicate possible leakage or wastage. Substantial savings are often possible if an adequate supply can be maintained using a smaller meter where charges are related to meter size. Note that remaining on an unmetered supply could be advantageous to some consumers. Major supplies can justify the recording of flow profiles when night-time consumption, for example, might indicate leakage or waste.

— Identify opportunities to reduce wastage in the form of leakage from pipes or fittings. Leaks may arise from minor individual sources, e.g. dripping taps or weeping glands, major leaks, with obvious visible signs, noise or pressure drop or medium-sized leaks, which could need specialised detection techniques. Large systems may benefit from zoning and pressure control.

— Identify opportunities to reduce wasteful use including tank overflows, unnecessary domestic use and excess process use. Commercial and domestic users can usually benefit from urinal controls, reduced toilet flush volume and restriction of wash basin taps. Process users may be able to employ water recovery techniques and benefit further from reduced water and effluent treatment costs. The need to maintain good housekeeping practices should also be recognised.

— Implement the available cost effective measures. Remember that reduced consumption is only of direct benefit if the supply is metered. A combination of conservation techniques and change of supply tariff may be required to achieve the best results.

— Monitor the results, ensuring that health and safety requirements are not compromised and that leak repairs have not simply increased system pressure and moved the problem elsewhere. Sub-metering should be considered for allocating costs on a departmental basis.

Bibliography

General

The Businessman's Energy Saver (Crewe: Formecon Publishing) (1989)

NIFES *Boiler Operators Handbook* (London: Graham & Trotman) (1989)

The Fuel Economy Handbook (London: Graham & Trotman) (1979)

Dryden I G C *The efficient use of energy* (Guildford: IPC Science and Technology Press) (1980)

Goodall P M *The efficient use of steam* (Guildford: IPC Science and Technology Press) (1980)

ENCODE (London: Department of Health/HMSO)

Audit guide on the management of energy in local authority buildings (London: The Audit Commission in collaboration with the Energy Efficiency Office)

NIFES *The Energy Manager's Handbook* (London: Graham & Trotman) (1979)

Helcke G A, Conti F, Daniotti B I and Peckham R J *The cost effectiveness of current building energy audits* (Luxembourg: Commission of the European Communities) (1990)

Source book for energy auditing (Stockholm: IEA/Swedish Council for Building Research) (1987)

Willoughby J *Energy conscious design for health care buildings principles and case studies* (York: University Institute of Advanced Architectural Studies) (1985)

Energy efficiency handbook **TR8** (London: Heating and Ventilating Contractors' Association) (1987)

CIBSE Publications

Environmental criteria for design *CIBSE Guide* Section A1 (London: Chartered Institution of Building Services Engineers) (1979)

Weather and solar data *CIBSE Guide* Section A2 (London: Chartered Institution of Building Services Engineers) (1982)

Thermal properties of building structures *CIBSE Guide* Section A3 (London: Chartered Institution of Building Services Engineers) (1980)

Air infiltration and natural ventilation *CIBSE Guide* Section A4 (London: Chartered Institution of Building Services Engineers) (1986)

Thermal response of buildings *CIBSE Guide* Section A5 (London: Chartered Institution of Building Services Engineers) (1979)

Internal heat gains *CIBSE Guide* Section A7 (London: Chartered Institution of Building Services Engineers) (1986)

Summertime temperatures in buildings *CIBSE Guide* Section A8 (London: Chartered Institution of Building Services Engineers) (1986)

Installation and equipment data *CIBSE Guide* Volume B (London: Chartered Institution of Building Services Engineers) (1986)

Reference data *CIBSE Guide* Volume C (London: Chartered Institution of Building Services Engineers) (1986)

Automatic controls and their implications for system design Applications Manual **AM1** (London: Chartered Institution of Building Services Engineers) (1985)

Guidance towards energy conserving operation Building Energy Code Part 3 (London: Chartered Institution of Building Services Engineers) (1979)

Measurement of energy consumption Building Energy Code Part 4 (London: Chartered Institution of Building Services Engineers) (1982)

CIBS Code for interior lighting (London: Chartered Institution of Building Services Engineers) (1984)

Appraising energy conservation investments using DCF methods Computer algorithm **CA1** (London: Chartered Institution of Building Services Engineers)

Department of Energy publications

Energy audits Fuel Efficiency Booklet **1** (London: Department of Energy) (Amended reprint 1989)

Steam Fuel Efficiency Booklet **2** (London: Department of Energy) (1985)

Economic use of fired space heaters for industry and commerce Fuel Efficiency Booklet **3** (London: Department of Energy) (Amended reprint 1990)

Compressed air and energy use Fuel Efficiency Booklet **4** (London: Department of Energy) (Amended reprint 1989)

Steam costs and fuel savings Fuel Efficiency Booklet **5** (London: Department of Energy) (In preparation)

Recovery of heat from condensate, flash steam and vapour Fuel Efficiency Booklet **6** (London: Department of Energy) (In preparation)

Degree days Fuel Efficiency Booklet **7** (London: Department of Energy) (Amended reprint 1989)

The economic thickness of insulation for hot pipes Fuel Efficiency Booklet **8** (London: Department of Energy) (Amended reprint 1990)

Economic use of electricity Fuel Efficiency Booklet **9** (London: Department of Energy) (Amended reprint 1990)

Controls and energy savings Fuel Efficiency Booklet **10** (London: Department of Energy) (Amended reprint 1990)

The economic use of refrigeration plant Fuel Efficiency Booklet **11** (London: Department of Energy) (Amended reprint 1990)

Energy management and good lighting practices Fuel Efficiency Booklet **12** (London: Department of Energy) (Amended reprint 1989)

The recovery of waste heat from industrial processes Fuel Efficiency Booklet **13** (London: Department of Energy) (Amended reprint 1990)

Economic use of oil-fired boiler plant Fuel Efficiency Booklet **14** (London: Department of Energy) (Amended reprint 1990)

Economic use of gas-fired boiler plant Fuel Efficiency Booklet **15** (London: Department of Energy) (Amended reprint 1990)

Economic thickness of insulation for existing industrial buildings Fuel Efficiency Booklet **16** (London: Department of Energy) (1985)

Economic use of coal-fired boiler plant Fuel Efficiency Booklet **17** (London: Department of Energy) (Amended reprint 1990)

Boiler blowdown Fuel Efficiency Booklet **18** (London: Department of Energy) (1983) (In course of revision)

Process plant insulation and fuel efficiency Fuel Efficiency Booklet **19** (London: Department of Energy) (Amended reprint 1990)

Energy efficiency in road transport Fuel Efficiency Booklet **20** (London: Department of Energy) (Amended reprint 1989)

Energy efficiency in buildings: How to bring down energy costs in schools (London: Department of Energy) (Second revised edition 1990)

Energy efficiency in buildings: Catering establishments (London: Department of Energy) (Second revised edition 1990)

Energy efficiency in buildings: Shops (London: Department of Energy) (Second revised edition 1990)

Energy efficiency in buildings: Further & higher education buildings (London: Department of Energy) (Second revised edition 1990)

Energy efficiency in buildings: Offices (London: Department of Energy) (Second revised edition 1990)

Energy efficiency in buildings: Sports centres (London: Department of Energy) (Second revised edition 1990)

Energy efficiency in buildings: Libraries, museums, art galleries and churches (London: Department of Energy) (Second revised edition 1990)

Energy efficiency in buildings: Hotels (London: Department of Energy) (Second revised edition 1990)

Energy efficiency in buildings: High street banks and agencies (London: Department of Energy) (Second revised edition 1990)

Energy efficiency in buildings: Entertainment (London: Department of Energy) (Second revised edition 1990)

Energy efficiency in buildings: Courts, depots and emergency services buildings (London: Department of Energy) (Second revised edition 1990)

Energy efficiency in buildings: Factories and warehouses (London: Department of Energy) (Second revised edition 1990)

Good Practice Guides (London: Energy Efficiency Office) (1989 onwards)

Index